BLACK LIGHT

A NOVEL THEORY OF THE UNIVERSE

BLACK LIGHT

A NOVEL THEORY OF THE UNIVERSE

by

Daniel Howitt

PARAGON HOUSE

First Edition 2021

Published in the United States by
Paragon House
St. Paul, Minnesota

www.ParagonHouse.com

Library of Congress Cataloging-in-Publication Data

Names: Howitt, Daniel, 1969- author.
Title: Black light : a novel theory of the universe / by Daniel Howitt.
Description: First edition. | Saint Paul, Minnesota : Paragon House, 2021. | Includes bibliographical references. | Summary: "This book reveals what the form of the structure of the universe is, and what the form of the content of the universe is, and how the prevailing accounts of these are inaccurate. Also revealed are the formal constraints on our ability to observe and conceive of what the structure and content of the universe are precisely. The novel observation and novel concept of black light, and the novel geometric analyses of what is necessary in order to observe and conceive of the intricacy of a geometric system, are used as the bases to critique the prevailing accounts of the form and content of the universe."-- Provided by publisher.
Identifiers: LCCN 2020036667 | ISBN 9781557789457 (paperback) | ISBN 9781610831277 (ebook)
Subjects: LCSH: Universe. | Dark matter (Astronomy) | Dark energy (Astronomy)
Classification: LCC QB791.3 .H69 2021 | DDC 523.01--dc23
LC record available at https://lccn.loc.gov/2020036667

Manufactured in the United States of America

10 9 8 7 6 5 4 3 2 1

The paper used in this publication meets the minimum requirements of American National Standard for Information Sciences—Permanence of Paper for Printed Library Materials, ANSIZ39.48-1984.

Contents

A Note from the Publisher

Throughout history, we have thought of darkness as the absence of light. Howitt argues that darkness is a different kind of light, which he calls "black light." He develops this novel proposition in the context of recent developments in physics and astrophysics related to black holes, dark matter, and dark energy.

By observing the motion of stars, astronomers since the 1920s have postulated that the universe is made up of more matter than can be seen with the naked eye. Dark matter is a different kind of matter that is described as not emitting, absorbing, or reflecting light. Dark matter is different from dead stars and dark nebulae because they reflect and absorb some types of light. Today physicists calculate that 95 percent of the universe is invisible, made up of 70 percent dark matter and 25 percent dark energy.

Howitt expands and challenges this developing notion of what lies behind and beyond the observable universe. To emphasize the formally different structure of the "external universe," he reconceives of dark energy as "black light," and dark matter "black matter." Black matter is what space, and all existents, are fundamentally comprised of, and is of a different nature, physics, and dimensional system than what our senses observe and our technology can detect. Not only is black matter a formally different kind of matter, but it emits black light, a formally different kind of light.

This new framework may provide superior and simpler answers to the problems of the geometric limits of the observable universe, what space is, what the fundamental constituents of the observable universe are, what gravity is, understanding quantum phenomena, the mechanism of cause of the observable universe, and other fundamental problems of astrophysics. Howitt does not merely lay out this theory abstractly, but describes how it relates to the theories of other physicists, the experiments at CERN, and the activities of space agencies.

This book suggests new areas of research for physics. It will also promote readers' creative thoughts about the universe and what lies behind and beyond the world we can observe.

—Gordon L. Anderson, Ph.D.
President, Paragon House

INTRODUCTION I-V

I

The prevailing concept of darkness is it is the absence of light. I demonstrate that darkness is actually a formally different kind of light (that is, light of a different nature, physics, and spatial dimensional system), which I refer to as "black light": The black of darkness is black light. I moreover demonstrate that it is emitted from "black matter," which is what space, and all existents, are fundamentally comprised of, and which also is of a different nature, physics, and dimensional system. (I reconceive of dark matter as black matter; and I reconceive of dark energy as black light). With this new understanding, the problems of whether there is a microlimit and macrolimit of the observable universe (and if so, what they are, and where they are), what space is, what the spatial dimensionality of the observable universe is, what the fundamental constituents of the observable universe are, what gravity is, what explains certain quantum phenomena, what the causal mechanism is that gave rise to the observable universe, and other fundamental problems, can be provided answers that are superior to the prevailing answers. Moreover—with this new understanding—other facets of the universe can be uncovered, such as the spatial dimensional system that underlies the observable universe, and its causal role, and mechanistic role.

II

I arrive at what I conclude after undergoing a multitude of developments and coetaneous revisions; and I elected to provide my early analyses, observations, conceptualization, and derivations in order to demonstrate how I arrived at my final analyses, observations, conceptualization, and derivations.

III

The primary questions that I attempt to accurately answer are the following: What is the form of the fundamental content of the universe, and what is it not?; What is the form of the fundamental structure of the universe, and what is it not?; and, What is the mechanism of cause of the universe, and what is it not? I construct answers to the aforementioned questions via my conception of, and observation of, black light, and via my particular geometric analyses: They allow for what I believe are accurate answers to the following secondary questions; and I believe that the answers can then be coalesced to accurately answer the aforementioned primary questions: Is there a microlimit and macrolimit of the universe?; Is the universe comprised of different spatial realms that are of different spatial dimensional systems?, What is the microlimit and macrolimit of the "observable universe"?; What is the microlimit and macrolimit of the "external universe"?; What is gravity?; Are there particles?; What is space?; What is dark matter?; What is dark energy?; What is time?; What is the spatial dimensionality of the observable universe?; What is the spatial dimensionality of the external universe?; What is the mechanism of the observable universe?; Where is the mechanism of the observable universe located?; What exceeds the observable universe, and where it is located?; and, What exceeds the micro external universe and macro external universe?

In my answers to the aforementioned questions, I demonstrate that an array of what are, by general consensus, conceived of as the fundamental questions of physics are actually not fundamental questions, because they are of inaccurate concepts, and inaccurate observations. The following questions are examples: What is the fundamental particle or particles of the universe?; What is the smallest kind of existent in the universe?; What caused the universe?; Is the universe is eternal?; Is the universe infinite?; What is the geometry of the universe?; Does the universe have a boundary?; Are there a multiplicity of universes?; Is the universe fundamentally composed of two-dimensional strings?; and, Is the universe is expanding?

The reason that I attempt to accurately answer what the "*form* of the fundamental content of the universe" is, and what the "*form* of the fundamental structure of the universe" is, is that, as I demonstrate, only the formal nature of both can be determined, because there are formal constraints on our ability to observe, measure, and conceive of what they are *precisely* via our observational capacity and conceptual capacity. Notwithstanding, I believe that my answers to the aforementioned primary questions and secondary questions entail demonstrating the extent to which prevailing physics, mathematics, astronomy, astrophysics, cosmology, the philosophy of physics, and the philosophy of cosmology are in error. Moreover, I believe that my answers provide the opportunity to attempt to develop perfectly justified methods of observation and experimentation which, as such, could possibly confirm what the form of the fundamental content of the universe is, and what the form of the fundamental structure of the universe is. The prevailing methods of observation and experimentation, as I demonstrate, cannot possibly confirm these. The methods reveal secondary features of the universe, rather than primary (or fundamental, or elementary) features of the universe. For example: Of what are conceived of "elementary particles," they are actually not separate existents, they are only relatively separate existents, and they are not fundamental (or elementary), nor are any other what I refer to as "relative particles." Moreover, the concept of elementary particle is, as I demonstrate, what I refer to as "conceptually dissociative," which is a type of severe conceptual error: What is dissociated from is that any relative particle, by virtue of being a relative particle, (a) consists of a sub-structure, regardless of whether there is ever the technology to observe or measure it, and (b) can be disintegrated, or divided, regardless of whether there is ever the technology to do so. Aside: As I argue, it would be when the disintegration or division of a relative particle yields a state of black light, rather than a sub-structure, that the fundamental substance of the observable universe is nearly reached, namely what I refer to as "black matter," which is formally different (that is, a different kind of matter, and of a different kind of physics) than anything of the observable universe. Black matter emits black light; and notwithstanding that black matter can never be observed, nor measured, due to being formally different, it can be more closely approached in the above way.

Regarding two of the aforementioned primary questions, what I mean by, "*fundamental structure* of the universe," is the system of spatial structures within which the content of the universe subsists, and which are, themselves, not structured by anything else, and what I mean by, "*fundamental content* of the universe," *is* the properties in the observable universe and external universe that are formally different than anything *of* the observable universe, and which, as such, cannot be precisely observed, or observed at all, and which cannot be precisely conceived of, and whose forms, as such, are all that can be conceived of. Moreover, such properties are the foundational properties with which the content of the observable universe is constructed, and upon which it is constructed.

IV

The darkness that is seen amid the illuminated, and the darkness that is seen in the absence of the illuminated, is itself illumination; and it is the only kind of illumination which, itself, can be seen. It is black light; and it is emitted from the content of the fifth spatial dimension of the observable universe, which is the three dimensional surface of the observable universe. The illumination that can override it for visual beings is commonly referred to as "visible light"; yet this kind of light is, itself, invisible, as it itself cannot be seen. It should therefore be referred to as "invisible light"; and black light should therefore be referred to as "visible light," and characterized as the only visible light.

If the observable universe was emptied of all invisible light, black light would be everywhere. It is, therefore, emitted from the content of the fifth spatial dimension of the observable universe, namely what I refer to as "black matter." And since the fifth spatial dimension is the microlimit of the observable universe, black matter forms the microlimit of the observable universe.

If the observable universe was emptied of all invisible light, and if there was no black light, there would be blindness, despite having the capacity to see. Blind persons do not see darkness, nor black.

Regardless of the intensity of invisible light in the observable universe, black light still pervades the fifth spatial dimension, including the fifth spatial dimension of the invisible light sources themselves. Ideal super-resolution microscopy would look beyond any intensity of invisible light, including on invisible light sources themselves, to black light.

While black light and black matter are *in* the observable universe, they are *of* the region of the universe that exceeds the microlimit and macrolimit of the observable universe, namely the micro external universe and macro external universe. They are therefore formally different than the content of the observable universe. While black light can be imprecisely visually observed, and while, as such, a semblance of black matter can be imprecisely visually observed, neither, due to being formally different than the content of the observable universe, can be measured, precisely observed, and precisely conceived of. That is, what they precisely are cannot be observed, measured, and conceived of.

The microlimit and macrolimit of the observable universe can be seen, though imprecisely, via the black light that they emit.

V

The following is an array of compressed discussions about some of the multitude of facets of the form of the fundamental content of the universe, and the form of the fundamental structure of the universe, that I observe, conceive of, and derive:

The universe consists of the observable universe, and the external universe. The observable universe has a microlimit and macrolimit. The microlimit is pervasive throughout the entirety of observable universe, including within all of the relative existents of the observable universe. The microlimit is what I refer to as "black matter." Moreover, the microlimit is what I refer to as the "three-dimensional surface" of the observable universe; and I refer to the three-dimensional surface as the "fifth spatial dimension." The observable universe is therefore what I refer to as "1-3|5 dimensional": the three spatial dimensions of length, width, and height, and the fifth spatial dimension. The, what I refer to as the "micro external universe," subsists within the entirety of the observable universe, and could be accessed from the observable universe via what I refer to as the "fourth spatial dimension," which I refer to as the dimension of "inward." The micro external universe is therefore what I refer to as "1-4 dimensional." The micro external universe, while within the observable universe, is not of the observable universe—that is, it is formally different. The dimension of inward is not the dimension of depth. The micro external universe consists of the mechanism of the observable universe. It is formally impossible to precisely observe, measure, and conceive of what the micro external universe is. It is moreover formally impossible for us to access the micro external universe. The fourth spatial dimension and fifth spatial dimension can be generally pictorially represented, and generally conceived of. Notwithstanding that of the five spatial dimensions, three (length, width, and height) can be precisely pictorially represented and precisely conceived of, and two (inward and three dimensional surface) can be generally pictorially represented and generally conceived of, they cannot exist in isolation from one another: The dimensions of the 1-4 dimensional micro external universe cannot exist in isolation from one another; and the dimensions of the 1-3|5 dimensional observable universe cannot exist in isolation from one another. As such, it is formally impossible for there to be one-dimensional existents, and two-dimensional existents. Moreover: Of the observable universe, and of the micro external universe, it is formally impossible for there to be 1-3 dimensional existents: The entirety of the three dimensionality of the observable universe includes the fifth dimension; and the entirety of the three dimensionality of the micro external universe includes the fourth dimension.

The micro external universe and macro external universe are of one realm, namely the external universe. As such, the macrolimit of the observable universe, while not of the fifth spatial dimension (that is, three-dimensional surface), is of the fourth spatial dimension. That is, it, unlike the fifth spatial dimension, is a boundary-limit, and is the same as any location of the fifth dimension, namely the access location to the fourth dimension of the 1-4 dimensional external universe. (The fifth spatial dimension is a three dimensional boundary limit; and while the macrolimit's boundary-limit is also 1-3|5 dimensional, it is less expansive as the microlimit: While it extends in the 1-3|5 dimensions, as is the case of the entirety of the observable universe, it extends in those dimensions to a considerably less extent).

Conceiving of any of the five spatial dimensions as existing in isolation from one another is to dissociate facets of the universe, that rely on one another for their existence, from one other. The ensuing mathematical and non-mathematical language that expresses the existence of the dimensions in isolation from one another is of what I refer to as "meaningless subtractive linguistic alteration."

At this stage of my discussion, I argue that there are no spatial dimensions that are beyond the five dimensions. Regarding the prevailing argument that there are spatial dimensions that are beyond the three spatial dimensions, and that they are "hidden" within what are referred to as "curled-up" existents, and, or, are curled-up themselves, the argument is incoherent, as (a) there are no even minimally precise pictorial representations, and minimally precise concepts, of such dimensions, (b) dimensions, themselves, cannot be curled-up, and (c) the description of curled-up existents as having such dimensions is meaningless: Nothing is conceptually expressed, nor pictorially expressed, by the description, as the concept of "curled-up existent" is meaningless, as is the concept of "curled-up dimension." Moreover, (d) arguing via mathematical language and non-mathematical language that there are "extra spatial dimensions," as they are referred to, because there must be in order for a theory to be coherent, but that they are hidden, is to engage in what I refer to as "meaningless additive linguistic alteration": Nothing conceptual, nor pictorial, is expressed, and instead, the language is psychological—that is, it arises from emotion, and is intended to elicit emotion. Moreover, arguing that the dimensions are "hidden" is done in order to evade the above problems (a) - (d).

The existence of the microlimit and macrolimit of the observable universe is determined by way of unaided vision and aided vision: The black that can be visually observed of both the relative space of the observable universe, and the existents of the observable universe, in the absence of, or the near absence of, common visible-light, is the microlimit; and the black that can be observed of the region of the macrolimit of the observable universe is the macrolimit. The black that is observed is black light; and black light is emitted from black matter. Both the microlimit and macrolimit of the observable universe are entirely comprised of black matter. Black light is the only kind of light which, itself, is visible, as it can be visually observed. What is commonly referred to as "visible-light" is invisible, as it itself cannot be visually observed. In spatial conditions that are, for humans, of darkness or near-darkness, if there was no black light, there would be blindness. Moreover, upon shielding one's eyes from visible-light, there would be blindness if there was no black light. The visual condition of darkness, or blackness, is different than the visual condition of blindness: Blind persons do not see blackness (or black-darkness).

If hypothetically there was no microlimit of the observable universe, there would be blindness in conditions of both black light and non-black light: Why this is the case for conditions of black light is explained above; and this is the case for conditions of non-black light because there would not be any color in the visual-field: All of the relative existents of the visual-field are fundamentally composed of the microlimit of the observable universe; in the absence or near-absence of black light, the existents visually appear as black or near-back, due to the expression of black light from black matter; in the presence of non-black light, the existents appear as various colors, due to reflection of the non-back light on the surfaces of the existents; in the absence of the microlimit of the observable universe, there would be no surfaces to reflect non-black light; the non-black light would, as such, pass through the visual-field without illuminating anything; and as such, there would be no color in thc visual field, including black (due to the absence of the microlimit); and with no color in the visual field, there would be blindness; color, including black, pervades the entirety of one's visual field. If, for this hypothetical, there hypothetically were no relative existents in the observable universe, except for a light source of any intensity that cannot, itself, be visually observed, there would be blindness even in presence of the non-black light.

If there was no macrolimit of the observable universe, the macro region of the observable universe would appear as a region of blindness instead of a region of blackness. The relative existents that are located between observers and the black macrolimit would be seen amid a matrix of background blindness.

That the observable universe is observable in the aforementioned two ways—that is, by way of its microlimit and macrolimit—entails that it has a geometry. In order to observe what the microlimit and macrolimit are themselves, it would be necessary to exceed those limits. That is, it would be necessary to be on the other sides of those limits. It is formally impossible to know where a limit is, and what a limit is, unless the limit is exceeded, and then observed, measured, and conceived of from that context. If one or more particular technologies seem to access and, or, exceed the limits of the observable universe, they have actually accessed and, or, exceeded sub-limits, and are, as such, observing additional features and realms of the observable universe. It cannot be determined a priori whether a technology can access the microlimit and macrolimit of the observable universe.

While the entirety of the form and content of the observable universe has not been observed, and is not understood, it, notwithstanding, has a general nature. If physical investigations of its micro region and macro region yield observations and, or, measurements of novel features, these features, by virtue of being observable, measurable, and conceivable, are still features of the observable universe. When physical investigations become unyielding, and remain unyielding, this may be indicative of how the microlimit and macrolimit have been nearly accessed. Again, while the microlimit and macrolimit are *in* the observable universe, they are not *of* the observable universe; and as such, they are formally different than anything of the observable universe; and as such, it would not be possible for human physical investigations to access them. It is when the physical investigations of the micro region and macro region of the observable universe yield the observation of black light, and continue to yield the observation of black light, that the microlimit and macrolimit have been nearly accessed. It is necessary to be formally different beings, and not of the observable universe, nor in the observable universe, in order to observe, measure, and accurately conceive of the microlimit and macrolimit. This is similarly the case regarding the observation of what the mind is itself: The mind cannot be used to observe what it itself is. It is necessary to exceed the mind, and have a formally different mind, in order to possibly observe what the mind is itself; and exceeding the mind, and having a formally different mind, would entail exceeding the observable universe, and no longer being in, nor of, the observable universe.

I briefly discuss the possibility that the micro region and macro region of the observable universe react in various ways to attempts to access the microlimit and macrolimit of the observable universe; but I do not develop this discussion, and instead, it is overridden by other discussions that I provide.

There are no macrolimits of the existents of the observable universe. If there were, there would be external universe between the existents. As such, there are no discrete existents of the observable universe, and instead, only relative existents. And as such, all of the relative existents are seamlessly integrated. And as such, the observable universe has the form of a liquid. Relative existents do, though, have relative macrolimits; and the relative macrolimits can be observed, and, or, measured.

Time is only the relationship between the duration of existence of the content of the observable universe, and mechanisms of regular accretional movement (and similar mechanisms), each movement of which is ascribed a numerical value. When time is conceived of as a property of the observable universe, the above relationship is dissociated from, and the general mechanism is conceived of as a property of the observable universe, rather than a mechanism in the observable universe. And as such, the dissociation entails the production of a delusive concept.

Time is only any time-mechanism *in* the observable universe; and conceiving of time as being a feature *of* the observable universe is to concurrently dissociate from what it is, and delusively, and incoherently, conceive of the observable universe as being a time-mechanism itself.

There is only relative space, namely regions of the spatial-field in which there is an absence of what are, for us, observable and, or, measurable relative existents. The microlimit of the observable universe pervades the entirety of the observable universe, including all of the relative existents of the observable universe.

In order to observe, measure, and accurately conceive of whether the observable universe is expanding, it would be necessary to exceed its macrolimit. That galaxies are moving outward and away from one another does not demonstrate that the observable universe is expanding, but rather, that the content of the observable universe is dispersing. "Dark energy" is actually black light; and it is black light that causes the dispersion of the content of the observable universe. The microlimit (which is what the macrolimit consists of), and, or, the micro external universe and the macro external universe, are the sources of black light.

Black light pervades the entirety of the observable universe, including in all conditions of 400nm-700nm light, and including on the relative surfaces of all 400nm-700nm light sources themselves, which entails that the fundamental color of the entirety of the content of the observable universe is black, and that the microlimit of the observable universe, which pervades the fifth dimension of the 1-3|5 dimensional observable universe, but which is of the 1-4 dimensional external universe, is black matter. Ideal super-resolution microscopy would show that black light pervades the entirety of the observable universe, including in all conditions of 400nm-700nm light, and including on the relative surfaces of all 400nm-700nm light sources themselves. The microscopy would, in a sense, look beyond the 400nm-700nm light, to black light.

Black holes are regions of relatively highly condensed black matter, and what I refer to as "black stars." The high condensation entails relatively extreme gravity, and the emission of an immense extent of black light.

Gravity, and different extents thereof, are phenomena of different extents of condensed black matter (that is, condensed microlimit). It is not possible for spatial regions to be curved, as they are simply extended without shape to their macrolimits. Only relative existents within spatial regions, and conglomerations of relative existents within spatial regions, can be curved; and they are curved only at their aforementioned relative macrolimits.

The regular spinning, rotation, and other regular movements of astronomical objects and phenomena are caused by the micro external universe: it is the mechanism of the observable universe. Black matter pervades all of the relative existents, and systems of relative existents, of the observable universe; and the micro external universe spins, rotates, etc., relative existents, and systems of relative existents, with black matter. Moreover, condensed black matter (that is, condensed microlimit) is what adheres systems of astronomical relative existents. Moreover, condensed black matter is the scaffolding within which galaxies form.

Gravity, then, is a combination of the aforementioned adhesive phenomenon of black matter (or condensation of black matter), and the mechanism of black matter. And as such, gravity is not a pull-down phenomenon, but rather, a push-down phenomenon; and it is the same in nature as the innate spin-phenomena, orbit-phenomena, rotation-phenomena, etc., of the micro observable universe and macro observable universe. And I will postulate that the different kinds of magnetism, and degrees of magnetism, are not pull-phenomena, but rather, push-phenomena that arise from the different kinds and degrees of magnetism increasing the speed at which black matter descends.

The purported concepts of zero, zero-dimensional, dimensionless, massless, colorless, nothingness, weightless, frictionless, and the like, lack conceptual content, are actually meaningless antonymic linguistic alterations, and only express and elicit emotion. As such, there are no such features of the universe.

The purported concept of infinity, as in infinite state, precludes its own coherence, as it can never be determined whether something is infinite in state. The purported concept of infinity, as in infinite process, is a conceptual dissociation, as, at any moment in the accretional procession of such things as physical division, mathematical division, physical addition, and mathematical addition, the procession is of a specific amount. To conceive of the procession as either occurring indefinitely, or as possibly occurring indefinitely, is to dissociate from the procession, and then use language that is of meaningless additive linguistic alteration.

The foundation of a substantial extent of mathematics is a multitude of dissociative concepts; and ensuingly, mathematics consists of a substantial extent of meaningless antonymic, subtractive, additive, combinatorial, separative, and other, linguistic alterations. Mathematics is founded by concepts, and is, as such, always secondary to the concepts; and as such, mathematics, independently, unless founded by accurate concepts, and, or, truly coherent concepts, will not only be incapable of addressing the matters of physics that I discuss, but will provide observations, theory, and statements that are inaccurate, incoherent, meaningless, delusive, dissociative, and psychological in nature.

Nima Arkani-Hamed and Jaroslav Trnka speculate that there is an object, which they refer to as the "amplituhedron," that is of the following nature; and their conception of the object is severely and multifacetedly dissociative: The object is of "pure geometry"; it does not have a content; it does not exist in time nor space; it does not exist in the universe; it is timeless (eternal); it is causal (it caused the universe via its pure geometry); and it is non-caused. Moreover, the mathematical and non-mathematical language that arise from the facets of the concept are, due to the dissociative nature of the concept, of the array of meaningless linguistic alterations. The mathematical and non-mathematical work of Arkani-Hamed and Trnka does not demonstrate that the concept of the object is even minimally coherent, nor, as such, that there could be such an object.

One of several final modifications that I undergo is that of deriving what I refer to as "absence," and arguing that it converts to what is opposite to it, namely the external universe and the observable universe. The external universe and the observable universe are caused via conversion from absence. We are, though, formally precluded from precisely conceiving of absence, and precisely conceiving of the mechanism of conversion. Absence is not a realm, nor relative existent, nor conglomeration of relative existents, nor state, nor phenomenon; and it as such is not observable, nor measurable. It can only be extremely generally conceived of, and primarily via what it is not, and via an analogy with a formal process of thought, namely the intermittent arising of thought via the conversion of absence to thought, and the intermittent discontinuation of thought via the conversion of thought to absence. The concept of the cause of the universe as being of conversion is superior to the concept of the cause of the universe as being of the interaction of preexisting relative existents or phenomena, or of the modification of one or more preexisting relative existents or phenomena. In each case, there is not anything that is truly caused.

Another final modification that I undergo is that I argue that the external universe (the micro external universe, and the macro external universe) has a microlimit and macrolimit, namely the sixth spatial dimension, which I refer to as "outward." Traversing those limits entails traversing to absence. To explain: To begin: As is the case of the observable universe, the microlimit of the 1-4 dimensional external universe is a three dimensional surface; and as such, both the observable universe and external universe have the fifth spatial dimension. (Aside, just as the macro limit of the observable universe is comprised of the micro limit of the observable universe, the macro limit of the external universe is comprised of the micro limit of the external universe; and as such, the macro limit of the external universe is also of the fifth spatial dimension). As such, the external universe is 1-4|5 dimensional. It is unclear, though, what the fifth spatial dimension of the external universe is comprised of, unlike what is the case for the observable universe: The fifth dimension of the observable universe is comprised of black matter, which emits black light. Notwithstanding, traversing the fifth dimension of the external universe would result in the conversion to absence. That is, the traverse of the fifth dimension of the microlimit and macrolimit of the external universe is a traverse to absence. The external universe is therefore 1-4|5|6 dimensional. Aside, the observable universe, which is 1-3|5 dimensional, is not 1-3|5|4 dimensional, because traversing its fifth dimension entails entry into the external universe via the external universe's fourth dimension (the dimension of inward). Traversing the fifth dimension of the external universe traverses to absence, via the sixth dimension (the dimension of outward); and since the sixth dimension is not a dimension of a spatial realm, it can be considered to be an aspect of the dimensional system of the external universe. Also aside: Absence, therefore, only partly converts to the universe.

The form of the content of the universe is the following: Absence; the mechanistic content of the external universe; black matter in the observable universe, which is *of* the external universe, and which is formally different in nature than the content of the observable universe, and which is of a different physics; the black light in the observable universe, which is of the external universe, and which is formally different in nature than the light of the observable universe, and which is of a different physics; the ordinary matter of the observable universe.

The form of the structure of the universe is the following: Absence; the 1-4|5|6 external universe; and the 1-3|5 observable universe.

While the precise content of the external universe and observable universe cannot be observed, measured, and conceived of, and while the fourth dimension of the 1-4|5|6 dimensional external universe cannot be entered by us from the 1-3|5 dimensional observable universe, and while the sixth dimension of the 1-4|5|6 dimensional external universe, which traverses to absence, cannot as such be traversed by us, we can observe black light, and we may be able to measure via microscopy, or one or more other methods, that black light pervades the entirety of the observable universe at the microlimit of the observable universe, including in all conditions of 400nm-700nm light, and including on the relative surfaces of all 400nm-700nm light sources themselves, which entails that the fundamental color of the entirety of the content of the observable universe is black, and that the microlimit of the observable universe, which pervades the fifth dimension of the 1-3|5 dimensional observable universe, but which is of the 1-4|5|6 dimensional external universe, is black matter, and emits black light. Moreover, we can observe the macrolimit of the observable universe via its black light. Moreover, we know that gravity is caused by condensations of the microlimit of the observable universe; and we know that the microlimit (space, which pervades the entirety of the observable universe) cannot be curved. And we know that a considerable extent of fundamental concepts of physics and mathematics are dissociative, and ensuingly of the many variations of meaningless linguistic alterations. And we know that the mechanism of the observable universe is the external universe, and that it, like the black light that it emits, is formally different than the observable universe. And we know that we, with our neurologies, and capabilities, are only able to observe, measure, and conceive of a semblance of the observable universe, and nothing of the external universe, except the black light that it emits into the observable universe. And we can derive the existence of the microlimit and macrolimit of the observable universe via the optics of black light, and via the geometric analyses that I provided. And we know that the observable universe is not fundamentally comprised of particles, nor any other kind of relative existent, and that it, instead, is of one contiguous substance. And we know that the external universe and observable universe were not caused via the modification of one or more preexisting relative existents, phenomena, or states, but rather, that they were converted from absence. And while we can only have extremely general concepts of absence, and the conversion of absence to presence, and mostly by way of the analogy of the conversion process of the mind, we do know the form of what occurs.

There are formal constraints that preclude us from observing, measuring, and conceiving of the precise fundamental content and structure of the external universe and observable universe, but via geometric analyses, the observation of a semblance of black matter, the observation of black light, optical geometric analyses, optical analyses, mathematical analyses, and novel conceptualization, we can conceive of their forms. Moreover, we can observe a facet of their forms via our observation of a semblance of black matter, and our observation of non-pure black light.

My initial geometric analyses are confirmed by what I later derive from my observation of black matter and black light.

If we could observe, measure, and conceive of the precise fundamental nature of the external universe and observable universe, we would not be as we are, nor where we are. As we are, and where we are, we can only possibly derive their forms. Although, we can observe a semblance of black matter; and we can observe non-pure black light; and we can possibly observe pure black light.

BLACK LIGHT

A NOVEL THEORY OF THE UNIVERSE

1. It is formally impossible to determine what *precisely* the microlimit and macrolimit of the universe are, because in order to do so it is necessary to exceed those limits: It is only from the perspective of what is beyond those limits, or on the other side of those limits, that those limits could possibly be precisely observed and, or, precisely measured.

2. It is impossible to determine if any observational technology or measurement technology is such that it has the capacity to determine what the microlimit and macrolimit of the universe are: It is only from a perspective that exceeds those limits that it could be assessed whether the technology succeeded at observing and, or, measuring those limits. Simply by virtue of a technology having elicited comparatively more fundamental features of the universe, and simply by virtue of the technology continuing to do so over time, does not entail that when the technology ceases to accomplish doing so—even upon the technology being improved in crucial ways—that the technology has succeeded in determining what the microlimit and macrolimit of the universe are.

3. Notwithstanding *(1)* and *(2)*, since the universe is the entirety of reality, it cannot have a microlimit and macrolimit, nor be unlimited: The presence of a microlimit and macrolimit entails that there is an other-side of those limits; and since the universe is the entirety of reality, there cannot be anything on the other sides of reality. Moreover, and as will be discussed, the concepts of unlimited, infinite, eternal, etc., are immensely dissociative, and, as such, incoherent. Moreover, there surely is an alternative to the states of being microlimited, macrolimited, and unlimited, and that it is something that is formally inaccessible to human experience and conceptualization.

4. There must be a microlimit and macrolimit of the *generally observable and generally measurable universe*, in light of how it is generally observable and generally measurable. (Hereafter, "generally observable and generally measurable universe" will be referred to as "observable universe"). As will be discussed, anything that is observable and, or, measurable is so because it is microlimited and macrolimited. Moreover, and as also will be discussed, the concepts of unlimited, infinite, eternal, etc., are immensely dissociative, and as such incoherent.

5. In light of how there must be a microlimit and macrolimit of the observable universe, there must be an external universe—that is, a universe that exceeds the microlimit and macrolimit of the observable universe: Anything that is microlimited and macrolimited is so because there is something that exceeds its microlimit and macrolimit: Hypothetically, if there was nothing that exceeds what is thought to be a geometric limit, there would be no basis to conceive of it as a limit.

If there is no proof, nor evidence, that there is anything that exceeds what is thought to be a geometric limit, and if, throughout the entirety of humanity, no proof, nor evidence, can be uncovered, this does not mean that the limit is a limit of the external universe.

It is of what formally cannot be observed, nor measured, and of what cannot be conceived of to be limited, and of what is not necessarily limited, that is a limit of reality.

Of what exceeds the microlimit and macrolimit of the observable universe, namely the external universe, it is surely neither limited nor unlimited: There surely is an alternative to the states of being limited and unlimited, and that it is something that is formally inaccessible to human experience and conceptualization.

6. Unlike what is the case of the observable universe, and as implied by *(1)*, and as is discussed from *(7)* onward, (a) nothing can be observed, nor measured, of the external universe, (b) and as such, nothing that is even minimally accurate can be conceived of the external universe, and (c) nothing that is even minimally coherent can be conceived of the external universe.

7. All conceptions of what exceeds the microlimit and macrolimit of the observable universe are inaccurate: All that can be accurately conceived of is that are such realms: It would be necessary for us to exceed the limits of the observable universe in order to have a basis with which it is possible to even minimally accurately conceptualize what such realms are.

8. In order for us to exceed the microlimit and macrolimit of the observable universe—which would entail that we physically exceed the macrolimit, and which would entail that we can exceed the microlimit by way of observation and, or, measurement—it would be necessary for us to become radically different in nature: We would not have any of the capacities with which to observe and, or, measure the observable and measurable intricacy of what was our observable universe, and instead, would only have capacities with which we can observe and, or, measure the observable and measurable intricacy of the realm that we are in, and, the fundamental intricacy of the observable universe that we left, such as its microlimit and macrolimit.

9. It is of course the case that any region of the observable universe is replete with the microlimit of the observable universe; yet, again, it is not only impossible for us to observe and, or, measure what it precisely is, we cannot with even minimal accuracy conceive of what it precisely is, except that it is black (as discussed below), three-dimensional, and pervades the observable universe.

10. That the observable universe is of course replete with its microlimit entails that the microlimit is of course a three-dimensional form.

11. In light of the aforementioned, it is likely the case that we are as we are—human—by virtue of the impossibility of accurately conceiving of, and precisely observing and, or, measuring (a) the microlimit and macrolimit of the observable universe, (b) the fundamental nature of the constituents of the observable universe, and (c) the external universe.

12. In light of *(11)*, and in light of how there is a microlimit and macrolimit of the observable universe, it is likely the case that attempts to measure the microlimit and macrolimit that approach succeeding at doing so result in the local inward expansion of the microlimit, and the local outer expansion of the macrolimit; and this would result in the continuation of our uncovering of what appear to us to either be comparatively more fundamental particles, fields, or other attributes, or what appear to us to be *the* fundamental particles, fields, or other attributes.

13. Black is of course a color; and the black that is observed of the general region of the macrolimit of the observable universe is, as such, caused by the presence of a substance: If the macro-region of the observable universe did not have a limit, then the blackness that is seen of it would instead be a region of blindness: There is a significant difference between what a sighted person sees in relatively purely dark conditions—namely blackness—and what a blind person observes everywhere.

14. The black that is seen of the macro-region of the observable universe is the macrolimit of the observable universe. Moreover, and as was discussed of the microlimit, any attempts that approach succeeding at measuring it would likely result in the outer expansion of it; and this would result in our continued uncovering of what appears to us to be either comparatively fundamental boundaries, or what appears to us to be *the* fundamental boundaries.

15. It is possibly the case that upon becoming significantly closer to the macrolimit of the observable universe, the semblance of it that can be seen appears to be of a different color, including of course white.

16. The totality of the microlimit of the observable universe cannot be a conglomeration of particles; and likewise, any relative point of the microlimit of the observable universe cannot consist of one particle: For anything to be observable or measurable, including of course a particle, it must be three-dimensional (that is, it must be extended in a region of relative space in all directions in order for it to be observable and measurable); moreover, such things can either be disintegrated with existing methods of disintegration, or could be so if methods are developed that could do so; and the fact that one or more particular methods do not succeed at doing so does not entail that the particle or particles that were attempted to be disintegrated are of the microlimit of the observable universe.

17. There cannot be zero-dimensional particles, because zero-dimensionality refers to the lack of existence. Likewise, zero is not a thing, nor a state of existence, but rather, the absence of anything; and while it is of course possible to discover the absence of something in particular, it is not possible to discover absence. Moreover, "zero" (zero itself), "nothingness," "weightless," "massless," "dimensionless," "frictionless," etc., are conceptually meaningless, antonymic linguistic alterations; and the alterations are undergone in order to elicit considerable emotion in oneself and others.

There cannot be massless particles, just as there cannot be empty glasses of water. "Massless particle," "dimensionless particle," etc., are the same as "no particle," or "the absence of a particle."

18. A particle of any nature and geometry, and any other feature of the observable universe of any nature and geometry, cannot have less than, nor more than, three spatial dimensions; and statements that something has zero spatial dimensions, and that something has one spatial dimension, and that something has two spatial dimensions, are conceptually meaningless, subtractive linguistic alterations; and a statement that something has more than three spatial dimensions is a conceptually meaningless, additive linguistic alteration; and both alterations are undergone in order to elicit considerable emotion in oneself and others.

Of some people, their concepts that certain things have one spatial dimension, and that certain things have two spatial dimensions, are, instead, disassociations from the concept of three-dimensionality: While three-dimensionality of course consists of three different dimensions, the dimensions are nevertheless dependent on one another for their existence, such that without any one of the three dimensions, none can exist; and to conceive of one-dimensionality, and two-dimensionality, as existing independently is to therefore disassociate facets of the universe that depend on one another from one another. Moreover, the way that such dimensionality is pictorially conceived of, and pictorially represented, is additionally disassociative. (a) A line—which of course is conceived of as one-dimensional—in addition to having length, has height, and depth: without depth, there would be nothing to rise from the spatial-field into existence; and without height, there would be nothing to extend across the spatial-field. Moreover, when a line is observed from each of its four sides, its depth can be observed; and to limit the observing of the line to only the top-view or bottom-view is negligent; and to conceive of the line as not being observable from any of its four sides is delusive. (b) Regarding a thing X that has a combined length and height, and which is conceived of as having only length and height (that is, without depth)—and it of course is conceived of as two-dimensional—there, again, would be nothing to rise from the spatial-field into existence; and to limit the observation to only the top-view or bottom-view is negligent; and to conceive of X as not being observable from any of its four sides, is delusive; and to conceive of X as being perfectly flat is incoherent, due to being a conceptually meaningless, subtractive linguistic alteration from the state of relative flatness. In both cases—*(a)* and *(b)*—the ineluctable existence of one or two other dimensions is disassociated from by way of negligent observation, and delusive pictorialization.

The concept that certain things have zero spatial dimensions, while a conceptually meaningless, but emotionally puissant, subtractive linguistic alteration, is not a disassociation from the concept of three-dimensionality: Three-dimensionality does not include the "zero-dimensional," and as such, nothing is disassociated from.

Aside: For something to exist in the observable universe, it must be spatial; and for something to be spatial, it must be dimensional; and since the "zero-dimensional" is not dimensional, there is no such thing as a "zero-dimensional" existent.

The way that "zero-dimensionality" is pictorially conceived of, and pictorially represented, while not disassociative of course, is negligent, and delusive: The point, or point-particle, of "zero-dimensionality" has depth, length, and height in the same way that was described of the "one-dimensional" and "two-dimensional"; and to limit the observation of the point or point-particle to only the top-view and bottom-view is negligent; and to conceive of the point or point-particle as not being observable from any of its four sides is delusive.

A drawn, or mathematically expressed, point, line, and collection of lines that form a geometry of some sort, are not existents that exist independently from the spatial world. Rather, they are as is described above, and they are misconceived of as explained above, and they are misobserved as explained above. At most, they could accurately be conceived of as relatively, minimally three-dimensional. Moreover, they are minimal representatives of objects of the typical observable universe; and such objects are of course three-dimensional.

The concept of certain existents as having more than three spatial dimensions does not of course entail dissociation of the concept of three-dimensionality, and instead is a conceptually meaningless, additive linguistic alteration; and since the alteration does not entail dissociation of the concept of three-dimensionality, it also does not entail the production of disassociative pictorializations.

The concept of a particular kind of existent as having, in addition to its "zero-dimensionalty," or "one-dimensionality," or "two-dimensionality," or even three-dimensionality, one or more other dimensions, does not entail that the existent has one or more additional spatial dimensions, but rather, that it has one or more other properties: Properties are most accurately described as properties, rather than as dimensions, as the description of dimensions is most apposite for spatiality. It would therefore only be accurate to describe what are argued to be additional dimensions as properties. And whereas dimensions are, of course, also properties, properties are not necessarily dimensions. Aside, an alternative to the above would perhaps be to describe what are argued to be additional dimensions as "non-spatial dimensions." In this case, it is clear that what is meant by "dimensions" is "properties." Also aside, perhaps the above is analogous to conceiving of a sixth sense, or an extra sensory faculty beyond the five sensory faculties: What is described to be the sixth sense, or extra sense, is not a sense, but rather, a property of the mind—that is, a property of thought, internal visualization, etc. The "sixth sense" would therefore be most accurately described as a non-sensory faculty, or a mental capacity. Therefore, conceiving of non-dimensional properties as extra dimensions, and non-sensory capacities as extra senses, is delusive; and publicly characterizing them as such is deceptive. Moreover, engaging in this conceptualization, and public characterization, is similar in nature to engaging in the previously discussed antonymic, subtractive, and additive linguistic alterations, and the previously discussed conceptual dissociation.

19. Time is (a) the duration-of-existence of something in relation to the origin of that thing; and the duration-of-existence can be measured by the consistent procession of a device that provides intermittent and accretional demarcations of the duration that the device has been proceeding, or (b) the post-origin measurement of the duration-of-existence of the thing with the device.

In the absence of the aforementioned device, other phenomena can be used, such as the phenomenon of the consistent procession of the appearance of the sun, and the phenomenon of the procession of thought, especially if the procession of the appearance of the sun, or the procession of the aforementioned device, are considerably internalized by the mind.

Conceiving of time as being an independent property entails dissociating from the relationship between the duration-of-existence of something and the devices that are used to provide measurements of its duration-of-existence; and conceiving of time as being something that is independent, and something within which things exist, is, as such, delusive: Can there be such a thing as temperature measurements that are independent of any thing?: Can it be X degrees independent of any feature of the observable universe?

Time is only the relationship between the duration of existence of the content of the observable universe, and mechanisms of regular accretional movement (and similar mechanisms), each movement of which is ascribed a numerical value. When time is conceived of as a property of the observable universe, the above relationship is dissociated from, and the general mechanism is conceived of as a property of the observable universe, rather than a mechanism in the observable universe. And as such, the dissociation entails the production of a delusive concept.

Time is only any time-mechanism *in* the observable universe; and conceiving of time as being a feature *of* the observable universe is to concurrently dissociate from what it is, and delusively, and incoherently, conceive of the observable universe as being a time-mechanism itself.

20. Even if a micro-particle was uncovered that could never be disintegrated, and whose content, as such, could never be analyzed, and even if a macro-boundary was reached that could never be penetrated, or disintegrated, this does not mean that the microlimit and macrolimit of the observable universe were found.

21. In order to observe and, or, measure what our minds precisely are, including their microlimits and macrolimits, surely this could only possibly be done by a being that is formally radically different than us: It is evident that the mind cannot be used in order to observe and, or, coherently conceive of what it itself is, just as the people of the observable universe cannot observe and, or, measure not only the microlimit and macrolimit of the observable universe, but the microlimit and macrolimit of any of the constituents of the observable universe.

Perhaps the aforementioned problem about the mind is not only an apposite analogy for the aforementioned discussion, and forthcoming discussion, about the observable universe, but perhaps it is the most apposite analogy: As is the case with the observable universe, in order for us to observe what our minds precisely are, it would be necessary for us to exit our minds; and that of course is formally impossible. And even of the hypothetical in which we can exit our minds, the hypothetical is too incoherent, because (a) the mind that is a feature of our brains—that is, the mind that we desire to precisely observe—would cease to exist, and (b) we would be no different than any mind that is external to our minds—that is, we would be no different than any other person. Moreover, even of the hypothetical in which we become a radically different kind of person, and as such have a radically different kind of mind, (a) we would no longer have the mind that we desired to precisely observe, and (b), and upon exiting our brains, the mind that is a feature of our brains would cease to exist. What would be necessary for our minds to even possibly be observed and, or, measured is the observation and, or, measurement by a radically different kind of being who is, moreover, outside of the observable universe, and not of the observable universe.

Would we rather, in order to resolve our curiosity about our minds, cease to be of our minds, cease to be the kind of beings that we are, and cease to be in the observable universe, or would we rather endure the humiliation of being incapable of having an even minimally accurate observation of and, or, measurement of, and concept of, what our own minds precisely are?

Similar to the aforementioned, would we rather, in order to resolve our curiosity about the observable universe, cease to be in the observable universe, and cease to be the kind of beings that we are?

It is astounding that we are formally constrained from observing, measuring, and even minimally coherently conceiving of some of the most fundamental, and seemingly simple, facets of the universe and ourselves. But, perhaps we are what we are, and as we are, because of this.

22. If I have a ten cubic foot tank of a particular liquid substance, and in the center of the tank released a conglomeration of relatively minute particles, which gradually dispersed, and so slowly that they would not reach the boundaries of the tank for approximately one hundred years, and if I ignored the existence of the boundaries of the tank, is my inference that the tank of water itself is expanding accurate?

23. It is impossible to even minimally accurately infer when the observable universe emerged, and how it emerged: The only way that such questions could possibly answered is from the perspective of what exceeds the microlimit and macrolimit of the observable universe, and, in particular, the perspective of beings who are formally different from us.

24. Aside, and a revision of *(8)*: In addition to how it is impossible to observe and, or, measure the microlimit of the observable universe, it is impossible to observe and, or, measure the macrolimit of any object of the observable universe, notwithstanding that we have exceeded their macrolimits. (It is of course the case that objects are objects by virtue of our having exceeded their macrolimits). All that we can observe and, or, measure is a semblance of their macrolimits, just as all that we can currently observe of the macrolimit of the observable universe is a semblance of it, and just as all that we could ever observe and, or, measure of it is a semblance of it. In order to observe and, or, measure the macrolimits of objects themselves, it would be necessary to exceed the microlimit of the observable universe: It could only be from that perspective that the macrolimits of the constituents of the observable universe could be observed and, or, measured, for from the perspective of what exceeds the macrolimit of the observable universe, it is likely the case that all that could be observed and, or, measured of the observable universe is its macrolimit. Moreover, that we exceed the macrolimits of the objects of the observable universe and still cannot observe nor measure the macrolimits themselves indicates that a further exceeding (the exceeding of the macrolimit of the observable universe) would not be useful, nor necessarily entail the acquisition of an additional observational and, or, measurement capacity with which the macrolimits of the objects of the observable universe could be observed and, or, measured.

Interestingly, by way of tactile perception, or the use of tactile instrumentation, we can indeed touch the macrolimits of the objects of the observable universe that we can touch. However, what we touch is not what we observe, nor what we measure. (Note: *(24)* is later revised and refined).

25. An additional revision of *(8)*: *(8)* is actually a nonsensical hypothetical, as not only is it impossible for us to observe and, or, measure the microlimit and macrolimit of the observable universe, it is, as such, impossible for us to physically exceed the microlimit and macrolimit of the observable universe; and as such, perhaps the hypothetical could have been better expressed as that if there were beings that existed in a realm that exceeds the observable universe, and if they could observe and, or, measure the macrolimit of the observable universe, they would be radically different than us.

If there were beings that existed in a realm that exceeds the observable universe, and if they were not able to observe and, or, measure the macrolimit of the observable universe, then they likely would be of the same, generalnature as us.

26. It is therefore evident that what is also necessary in order to observe and, or, measure the macrolimit of the observable universe is, in addition to a perspective that exceeds the macrolimit of the observable universe, one that is formally different than our general perspective: Being on the other side of the macrolimit of the observable universe does not necessarily entail that the limit could be observed and, or, measured.

27. An addition to *(24)*, and as explained further in subsequent discussions: Analogous to what was discussed of the macrolimit of the observable universe, if there was no microlimit of the observable universe, we, despite having the capacity to see, would be blind: Analogous to what was discussed of the macrolimit, the various colors of objects that we see, including of course black and white, is a semblance of our microlimit, and that without a microlimit, there would be no color, and instead, blindness.

The ever-changing macro-boundaries of our visual-fields are replete with color, because all of the objects of our visual-fields are colored. In the absence of such color, we would be blind, despite having the capacity to see.

In conditions of the partial supplanting of the black light of our visual-fields with what, for us, is invisible light (namely, and approximately so, 400nm-700nm light, which, itself, is invisible for us), there is blindness between our eyes and the content of our visual-fields. This blindness enables us to see the ever-changing, colored, macro-boundaries of our visual-field. If the colored macro-boundaries were removed, there would be complete blindness.

28. In addition to how it is surely the case that the microlimit of the observable universe could not be observed and, or, measured from the perspective of what exceeds the macrolimit, it would seemingly be impossible for there to exist an observational perspective and, or, a measurement-capable perspective in what exceeds the microlimit, in light of its relative minuteness. It therefore can seemingly only be hypothesized that if it was possible for there to be an observational perspective and, or, a measurement-capable perspective in what exceeds the microlimit, the microlimit could be observed and, or, measured, and the macrolimits of the constituents of the observable universe could be observed and, or, measured. Moreover, and perhaps analogous to how it would seemingly be impossible to observe and, or, measure the microlimit of the observable universe from the perspective that exceeds the macrolimit, it would surely be impossible for the macrolimit of the observable universe to be observed and, or, measured from what exceeds the microlimit of the observable universe: It is likely the case that on the other sides of both limits, all that could be observed and, or, measured is the limit that one is on the other side of. And for this reason, I should retract my previous discussion that the only perspective from which the macrolimits of the constituents of the observable universe could be observed and, or, measured is that of what exceeds the microlimit. (Note: *(28)* is later revised and refined).

29. The content of the observable universe of course consists of extraordinary physical complexity; and much of the content seems to be extraordinarily observationally elusive and, or, inexplicable. But, the observable and, or, measurable content of the observable universe seems to be secondary in importance: While it is of course important to observe and, or, measure, and describe, and explain, such content, it seems that doing so is secondary in importance to the fundamental issues of what are, and where are, the microlimit and macrolimit of the observable universe, and where, when, and how the observable universe emerged, or began to emerge.

30. The microlimit and macrolimit of the observable universe cannot be fields for the reasons that are mentioned in *(16), (18)*, and especially *(108)* onward.

31. In addition to what was previously discussed about the inward expansion of the microlimit, and the outward expansion of the macrolimit, it is likely the case that the inward expansion of the microlimit entails the creation of comparatively more fundamental particles, which may emerge as sub-structures of disintegrated particles.

In light of how the region of the macrolimit has not yet been experimentally investigated as the apparent region of the microlimit has, and in light of how it may essentially always remain quiescent due to this, there of course is uncertainty about what would occur upon succeeding at being within close proximity of it. I would surmise that it would expand outward, and that it would continue to do so in response to all of such approaches.

32. Of course the observable universe is replete with the microlimit; and as such, the microlimit extends to the macrolimit; and as such, the macrolimit is where the microlimit discontinues.

33. In light of the aforementioned discussions about the limits, it is surely the case that what exceeds the limits is formally different than the observable universe: If the relative micro-realm and macro-realm that exceed the limits of the observable universe were of the same formal nature as the observable universe, then our microlimit and macrolimit (a) would be accessible, (b) comprised of observable and, or, measurable attributes, (c) and shared between the other realms such that what would actually be the case is that there are not three, discreet realms, but rather, one realm.

Notwithstanding the aforementioned, there is no basis to think that what exceeds the microlimit is a micro-universe of the same, formal nature as ours, and that what exceeds the macrolimit is a macro-universe of the same, formal nature as ours.

34. It is likely the case that what exceeds the microlimit and macrolimit are combined, and, moreover, that what exceeds the microlimit is what provided, and perhaps continues to provide, the constituents of our universe, and that what exceeds the macrolimit is what provided, and perhaps continues to provide, the space in which the constituents reside.

35. Based on the aforementioned discussions, while it is of course the case that the observable universe has either a general geometry, or an ever changing, general geometry, or a chaotic geometry, or an ever changing, chaotic geometry, it is also of course the case that it is impossible to observe and, or, measure it.

36. "Infinity" and "zero" (zero itself, rather than the absence of X) are typically conceived of as states of existence; but in order to observe and, or, measure those states, it would be necessary to exceed the negative limit of zero, and to exceed the positive limit of infinity; and doing so would be precluded by the states themselves: Without having exceeded the negative limit of zero—that is, the side of the limit of zero that is opposite to 1, or opposite to X—and without having exceeded the positive limit of infinity—that is, the side of the limit of infinity that is opposite to infinity minus 1, or infinity minus X—it would not be possible to assess whether the state of zero exists, and whether the state of infinity exists; and since, according to the general concepts of zero and infinity, there is nothing less than zero, and nothing more than infinity, there would be no possibility of exceeding the negative limit of zero, nor the positive limit of infinity.

The general concepts of infinity and zero themselves demonstrate the impossibility of the states of existence of infinity and zero; and the general concepts of infinity and zero demonstrate their own incoherence.

37. Such a property as space seems to exist.

If there was not space itself, then the constituents of the observable universe would be permanently fixed in their locations.

The objects of the observable universe do not displace space, or replace space, because that would entail that the objects are permanently fixed in their locations.

That objects seem to exist *in* space, and move *through* it, does not entail that they displace it, nor replace it, for the aforementioned reason.

It must therefore be the case that space is within objects, and that the presence of objects does not alter it in any way.

In order to observe and, or, measure what space is, it would be necessary to exceed space—that is, it would be necessary to exceed either the microlimit of space, or the macrolimit of space—and as discussed previously, that is formally impossible.

We can observe a semblance of space by way of our unaided vision, and by way of our aided vision, namely, the regions of the relative absence of objects that are between the objects that we see. Moreover, we can experience a semblance of space by way of our existence, and by way of our movement: As spatial objects, we exist within space, and can experience ourselves as such; and when we move, we can experience ourselves moving within space. But, the aforementioned is not space itself, just as the semblance of the macrolimit of the observable universe that we can observe, and the semblance of the macrolimits of constituents of the observable universe that we can observe, are not macrolimits themselves.

38. The goal should therefore not be to uncover the microlimit and macrolimit of the observable universe, but rather, to exceed the micro-realm of the observable universe, and to exceed the macro-realm of the observable universe. That is, the goal should therefore not be to uncover the fundamental micro-constituent or micro-constituents of the observable universe, but rather, to exceed that realm of the observable universe; and the goal should therefore not be to uncover the macro-boundary of the observable universe, but rather, to exceed that realm of the observable universe.

The aforementioned goal does not entail the goal of uncovering the microlimit and macrolimit of the observable universe, but rather, ignores that goal.

Notwithstanding that it is formally impossible to accomplish the aforementioned goal, it at least is the goal which, if accomplished, would entail at least the possibility that the microlimit and macrolimit could be observed and, or, measured.

39. The partial disintegration of protons by way of colliding them with one another at a relatively extreme speed would seemingly not possibly accomplish the goal of exceeding the microlimit, regardless of the speed at which the protons collide with one another. It seems that the most relevant experimentation, or perhaps the only relevant experimentation, for the purpose of attempting to accomplish the goal is (a) the use of explosive technology in perhaps any region of space, or a particular kind of region of space, and the observation and measurement of not what occurs in the explosion itself, but rather, what occurs to the surrounding region of space, and (b) the use of object projection technology (or object acceleration technology) in order to project or accelerate an object or objects against any region of space, or a particular kind of region of space, and the measurement of not what occurs to the object or objects, but rather, what occurs to the space that is being subjected to the object or objects.

Regarding particle collision experimentation, what would seem to be most relevant is not what occurs to the particles that are collided, but rather, to the surrounding region of space.

It is likely the case, though, that of the experimentation of the aforementioned methods that approaches being successful at exceeding the microlimit of the observable universe, what would result is either (a) the inward expansion of the microlimit, or (b), *(a)* and the creation of comparatively more fundamental particles. (This portion of *(39)* is later revised and refined).

40. It is impossible to conceive of a particle as being indivisible (or disintegrable): A particle, by virtue of it being a particle—that is, a three-dimensional form—can always be conceived of as being disintegrable, regardless of whether the practical capacity exists to disintegrate the particular kind of particle that is conceived of. Moreover, conceiving of a particle as not consisting of any content (or structure) is actually a conceptually meaningless, subtractive linguistic alteration; and it is a conceptually meaningless subtractive linguistic alteration of the kind that was discussed earlier; and it moreover is tantamount to conceiving of a particle-less particle.

Likewise, of any particle that is uncovered, it, by virtue of being a particle, could possibly be disintegrated, regardless of whether the capacity exists to disintegrate it.

In light of how it is of course the case that the microlimit of the observable universe cannot be a conglomeration of particles, it is possibly the case that particle disintegration that approaches the microlimit of the observable universe entails the creation of one or more sub-particles of a more fundamental nature in the particle or particles that are being disintegrated; and the creation, of course, would entail that the particle or particles that are being disintegrated are conferred one or more sub-particles from what exceeds, or is on the other side of, the microlimit of the observable universe.

As has been the case for considerable time, and as will likely be the case for considerable time in the future, and possibly for as long as particle disintegration experimentation continues: Of the particles that are uncovered and conceived of as being the most fundamental particles of the universe, they will eventually be disintegrated into comparatively more fundamental particles; and it has been, and would continue to be, impossible to observe and, or, measure if the sub-particle or sub-particles that were uncovered were already a part of the structure of the previously uncovered particle or particles, or whether they were created (that is, conferred) during the particle disintegration experimentation itself.

It could of course be the case that all of the particles of the observable universe already consist of such an extraordinary accretion of sub, sub-sub, sub-sub-sub, etc., particles that particle disintegration experimentation will never, for the entirety of the particle disintegration experimentation of humanity, entail the creation of, or conferring of, sub-particles.

41. Since the macrolimit of the observable universe is not ubiquitous, as is the case of the microlimit, and since our observation, measurement, and exploration of the macro-region of the observable universe has never yielded the presence of what seems to be a region that is even approximately close to the macrolimit, the aforementioned methods, namely *(a)* and *(b)* of *(39)*, would not even possibly be effective in any way.

In case our observation, measurement, and exploration of the macro-region of the observable universe yields uncovering the presence of what seems to be a region that is approximately close to the macrolimit—for example, perhaps what appears to be an encasing boundary of the observable universe is uncovered to exist at what, at that time, is a relatively accessible distance away—and in case our observation, measurement, and exploration of the macro-region of the observable universe yields accessing what appears to be an encasing of the observable universe, it is not necessarily the case that what seems to be a region that is approximately close to the macrolimit is so, and that what appears to encase the observable universe does so. For example, even if what seems to be an encasing boundary is uncovered, and even if, after 1,000 years of subjecting regions of it to ever advancing, disintegrating technology, penetrating technology, imaging technology, and perhaps other kinds of technology, the regions of it still remain integrated, unpenetrated, and unimaged, this does not entail that the macrolimit of the observable universe was uncovered, because it is necessary to exceed the macrolimit of the observable universe in order to uncover where it is, and what it is. Likewise, if one or more regions of the aforementioned encasing boundary are disintegrated, or penetrated, or imaged, such that the negative side (the other side) of the apparent, macrolimit is observed and, or, measured, and even if, on the side of the apparent limit, there are no observable and, or, measurable objects, this still does not entail that the macrolimit of the observable universe was uncovered, and actually entails that it was not uncovered.

42. To some extent akin to what was discussed about the macrolimit of the observable universe, if hypothetically there was no microlimit of the observable universe, but still a sufficient light source, we, notwithstanding that we have the capacity to see, would be blind, because (a) there would be no substance in the observable universe, and instead, only space, and (b) the space would be a region of blindness: Without a microlimit, there would be no substance in the observable universe, and instead, only space; and with the observable universe consisting only of space, there would be no color; and without the presence of color, including white and black of course, we would be blind, notwithstanding that we have the capacity to see.

Of course the aforementioned hypothetical is immensely dissociative in nature, as it dissociates from how we, and a light source, would not exist according to an aspect of it itself; but of course it is the case that hypotheticals can be used in this way in order to consider what would be the case in circumstances that have not occurred, and in circumstances that could likely never occur, and in circumstances that could never occur.

I was considering revising the aforementioned hypothetical by stating that due to the presence of the macrolimit, we, instead of being blind, would see black; but in light of how the space between us and the macrolimit would be a region of blindness, we would not see the black semblance of the macrolimit, even if, hypothetically, the macrolimit was, for example, one millimeter in front of our eyes.

43. "Infinite," in the sense of "infinite process," which I suppose is synonymous with "eternal process"—for example, the infinite division of particles, the infinite serial accretion of numbers, etc—is a dissociative and antonymic concept: (a) In order to know whether processes *do* occur infinitely, it of course is necessary to observe and, or, measure that that is the case; and that is of course impossible; (b) stating that processes *could* occur infinitely is false, because what is necessary in order to ascertain whether that is the case is that the processes *are* conducted infinitely; and that of course is impossible; and (c) stating that processes *could* occur infinitely *if* they proceeded infinitely is merely hypothetical. The concept is dissociative and antonymic in the sense that it entails that the limit—namely, what *does* occur—is dissociated from, and then antonymically expressed to occur without limit; and the antonymic expression is a conceptually meaningless, but an intensely, emotionally eliciting, linguistic alteration.

It is the dissociation from the limit that entails the eliciting of immense emotion; and the private use, and public use, of the antonymic expression elicits the dissociation, and concurrent emotion, in oneself and others. That is, "infinite," "eternal," "limitless," etc., are emotionally puissant because they entail the aforementioned dissociation.

Conceptual dissociation, including if what ensues is the absence of alternative conceptualization, can entail the arising of immense emotion. Likewise, psychological dissociation can entail the arising of immense emotion.

Conceptual dissociation is typically undergone in order to supersede one or more undesirable conceptual states, and in order to acquire a comparatively more preferable conceptual state and emotional state; and psychological dissociation is typically undergone in order to supplant the memory of, and, or, the psychological effects of, past psychologically deleterious phenomena, and, or, the awareness of present psychologically deleterious phenomena, and in order to, as such, acquire a comparatively more preferable conceptual state and emotional state.

44. In light of how, according to the concepts of "infinity," "infinite process," "eternity," etc., there is no limit of "infinity," "infinite process," "eternity," etc., it is of course not possible to observe and, or, measure them by exceeding them; and as such, the concepts themselves preclude the possibility of determining whether there are such properties and processes in our universe.

The aforementioned concepts are not delusive, but rather, incoherent.

45. Dr. Robbert Dijkgraaf states the following:

> One of the most amazing things we discovered in science is that everything is made of small particles. ... But how small are these particles? ... And does the search ever stop?[1]

> But we kind of know that in terms of if you go shorter, shorter, shorter, that at some point at this very very end ... you can't really go smaller than that: that's kind of the smallest distance there in nature. So at some point it has to stop.[2]

By "smallest distance there in nature," he of course is referring to the nature of "the smallest particles," and expresses his belief that an aspect of the nature of the smallest particles that constitute every thing of the universe is that they are of such a small size that it is not possible for there to be anything smaller: "you can't really go smaller than that"; and, "that's kind of the smallest distance there in nature." He moreover believes, regarding particle experimentation, that "at some point it has to stop," and that when it does, it will have revealed the "very

1. The Institute for Advanced Study, Article: "The Smallest Particles: What Do They Reveal?," https://www.ias.edu/ideas/2013/dijkgraaf-particles-video, November 16, 2013.

2. The Institute for Advanced Study, Video: "The Smallest Particles: What Do They Reveal?," 50:50-51:20, https://www.ias.edu/ideas/2013/dijkgraaf-particles-video, November 16, 2013.

very end," namely, "the smallest distance there in nature." He therefore seems to believe that once particle experimentation is not capable of revealing particles that are smaller than the previously revealed ones—that is, once particle experimentation, despite being proceeded with, comes to a "stop"—the "very very end," that is, "the smallest distance there in nature," will have been reached.

The problems with the aforementioned were discussed previously.

46. It is possible that the previously discussed macrolimit is simply an intermediary macrolimit; and if, in the future, it is thoroughly observed and, or, measured, then it surely is.

47. An addition to *(43)*: Hypothetically, if the observable universe did not have a microlimit—and as such, no objects, and no light sources—and no macrolimit, a person, notwithstanding that he or she has the capacity to see, would be blind: He or she would not see blackness, nor a black region of space, but rather, would be blind.

With our eyes open while we are in an essentially completely dark room, if the observable universe did not have a microlimit, we would not see blackness, nor a black region of space, but rather, would be blind. And if a microlimit appeared, we would see blackness, or a black region of space, and would notice the difference between having been blind and being able to see the black region of space.

48. While, akin to how the observable universe must have a microlimit and macrolimit, it of course had an origin, and may, likewise, have a finality. Moreover, it may have another origin, and another finality, and so on. Moreover, of the current observable universe, it may not be the original observable universe. In contrast, and akin to how the external universe cannot have a microlimit, nor macrolimit, nor be unlimited, it (the external universe) also cannot have an origin, nor be "infinite." Moreover, surely there is an alternative to the state of origination, and that it is something that is formally inaccessible to human experience and conceptualization.

Since the origin of the observable universe already occurred, it of course cannot be observed, nor accurately conceived of, and only speculated about.

Since the origin of the observable universe surely consisted of creation, or origination, from that which exceeds its microlimit, and from that which exceeds its macrolimit, it is formally impossible to accurately conceive of even the general form of the creation or origination.

Of whatever is coherently conceived of—of the origination of the observable universe—it is secondary to the actual origination; and the fact that something is coherently conceived of demonstrates that it is secondary.

It would be necessary to be outside of the microlimit and macrolimit of the observable universe—that is, outside of the observable universe—in order to possibly observe, and, or, measure, and as such know, or only coherently conceptualize, the origination of the observable universe; and in light of our observational, measurement, and conceptual incapacities, we surely would have to be formally different than we are in order to even possibly be able to observe, and, or, measure, and as such know, or only coherently conceptualize, the origination of the observable universe.

If we could coherently conceptualize the origin of the observable universe, and coherently conceptualize (or observe, and, or, measure, and as such know) the microlimit and macrolimit of the observable universe, and coherently conceptualize (or observe, and, or, measure, and as such know) the microlimit and macrolimit of the constituents of the observable universe, we would be formally different in nature than we are, and—moreover—not of the observable universe, nor a part of it, nor—as such—able to participate it.

49. There can be only one universe of course—that is, there can be only one reality—but there may be one or more observable universes, because the observable universes could be separated by external universe.

50. Whereas the macrolimits of the constituents of the observable universe are of course different than the macrolimit of the observable universe, the microlimits of the constituents of the observable universe are the same as the microlimit of the observable universe.

51. The microlimit of the observable universe is therefore likely space itself. And just as any even partly observable and, or, measurable object or realm must have a microlimit and macrolimit, and just as there are two sides of any spatial limit, there is an other side of the microlimit of the observable universe that is opposite to the side of the microlimit where the observable universe is located; and it is only from that side of the limit that the limit could possibly be observed and, or, measured.

52. In light of how it is impossible for us to observe and, or, measure, and as such, even minimally coherently conceive of, what space is itself, no arguments can be true according to which space, in the past, had one or more properties, and, or, currently has one or more properties, and, or, must have one or more properties, and, or, may in the future have one or more properties, such as a general size, general date of origin, general date of termination, and general geometry, and curvature, expansion, disintegration, diminution, compression, etc. All that can accurately be observed, and as such, conceived of, is that space itself is black, that it pervades the observable universe, and is three-dimensional.

53. If we could observe and, or, measure space itself, we would, ignoring for the moment that we would be a formally different kind of being, and not of the observable universe, and not a part of the observable universe, be incapable of observing the observable universe.

54. Space cannot not be curved, except at its macrolimit. (a) The interior of any region of something, regardless of the extent to which its macrolimit is curved, cannot be curved. Rather, the interior, itself, is simply extended to the macrolimit without any shape. (b) That particular kinds of phenomena occur within the interior of something that seemingly could only occur if the interior is curved does not mean that the interior is curved, but rather, that the interior consists of constituents and phenomena, which, when they interact with each other, or with something else that enters into the interior, give the appearance that the interior itself is curved. For example: A region of water, akin to the interior of any region of something, cannot not be curved, except at its macrolimit; when various kinds of things interact with the water in various ways, and the result of the interaction is such that it seems that the water itself is curved, what of course actually occurs is that various constituents of the water, and, or, various phenomena of the water, interact with the things in such a way that it seems as if the water itself is curved. (c) As an elaboration of *(b)*: The interior of a region of something cannot, by virtue of being a three-dimensional region that is extended to its macrolimit, be curved. Rather, the interior is simply extended three-dimensionally to the macrolimit without any shape. In contrast, the constituents and phenomena of the interior region of something can, in a sense, be curved: There may be regions of collections of constituents that are curved; and there may be regions of phenomena that are curved. For the aforementioned water example, if a beam of light, or a stream of a substance, is sent into a region of water in which there is a curved current, or a whirlpool, or a collection of constituents that are curved, and if the light or substance curve with these phenomena and constituents, this of course does not mean that the water itself is curved. Moreover, the beam of light, and the stream of substance, even in the absence of the aforementioned constituents and phenomena, would, due to slowing in speed, due to traveling through a region that consists of constituents and phenomena, begin to curve, until they finally discontinue both traveling and existing. This, too, does not mean that the water itself is curved.

Pictorializations and physical demonstrations of "curved space," or "warped space," are prevalent, and consist of one or more macro-boundaries of a region of pictorialized space, or a physical form, being curved in various ways. The space, or physical form, underlying the curved macro-boundaries is not curved. Moreover, notwithstanding that such pictorializations and physical demonstrations are of the macro-boundaries of the pictorializations and forms, this is dissociated from, as it is then meaninglessly stated that the interior of the entirety of the displayed space, or the interior of the entirety of the displayed physical form, are also curved: This is meaningless because when this is stated, nothing is conceived of. That is, this is a purely non-conceptual, linguistic statement, and is akin to the aforementioned linguistic alterations, namely the antonymic, additive, and subtractive linguistic alterations. Aside, that something can be stated, including with mathematical language, does not necessarily entail that what is stated is conceptually meaningful.

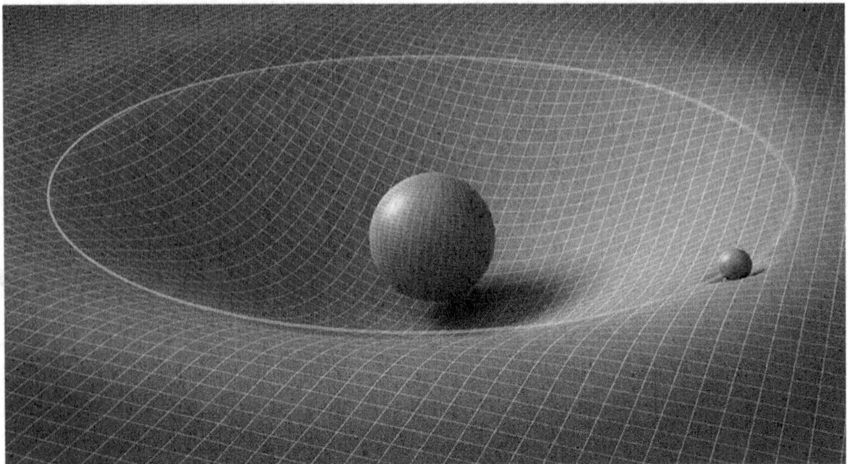

Moreover—for the above pictorialized space and physical forms—only a slice of pictorialized space is provided, and only a slice of physical form is provided; and then the top surface of the slice, along with the underlying space or physical form, is displayed. The top surfaces, which are the top macrolimits of the slices, are displayed as being curved in various ways, but usually with a half-spherical concave indention in a central region, which is stated to be caused by the presence of a spherical mass, which, for the above pictorializations, is usually shown hovering immediately over the concave indentions; and for the above physical forms, often a spherical mass is placed over the physical forms, which are such that the weight of the masses causes the spherical concave indentions in the physical forms. (a) With regard to the pictorializations, only one concave indention is displayed in most pictorializations; and when such indentations are placed around the entirety of the sphere, the result is simply a larger sphere around the sphere, and slightly outside of the sphere:

Since the pictorializations that show one indentation usually do so with a grid-pattern at the macrolimit of the region of space that is shown, and, by distending the grid-pattern at the bottom of the sphere, the larger sphere around the sphere would consist of distended grid-squares; and the spatial region outside of the distended grid-squares would consist of grid-squares that are not distended:

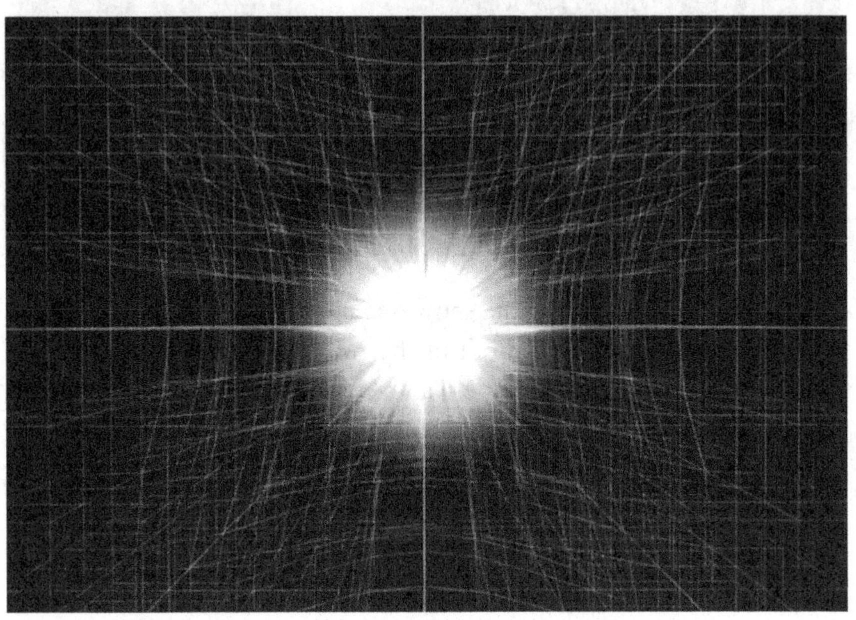

However, since space is a three-dimensional form, only its macrolimit could be curved, as is the case with liquids, solids, and collections of gasses. While space, as is the case with liquids, solids, and collections of gasses, may be comparatively more or less condensed in various regions, it cannot be curved, bent, warped, distended, etc., just as it is not possible to curve interior regions of liquids, solids, and collections of gasses. The likely most accurate explanation, and perhaps the only non-dissociative explanation, is that space is more condensed in the regions surrounding relatively discrete physical existents, and less condensed elsewhere. And regarding dissociation: (b) With regard to the above discussed illustration of curved space via the use of a spherical mass that is placed over a physical form which is such that the weight of the mass causes a spherical concave indention in the physical form, this phenomenon, which is limited to the use of masses of particular weights, and physical forms of particular natures, is dissociated from, and merely assumed to occur between discrete physical existents and space. But more importantly, and similar to what was discussed above in *(a)*, if a spherical mass is implanted in the middle of a physical form, it does not bend any region of the physical form, but rather, displaces some of it, or compresses some of it: The form, around its entire macrolimit, may distend outward slightly, due to the spherical mass occupying a region where a portion of its form was; or the form may become more condensed throughout, though especially so in the area adjacent to the spherical mass. And if the form, around its macrolimit, distends outward slightly, this does not mean that the form that is underlying the macrolimit has curved, but rather, only that the macrolimit has curved (or that is, that is has been curved).

55. In light of how there is not a state of "nothingness," etc., and in light of how space pervades the entirety of the observable universe, space is therefore a physical property. And in light of how space is a physical property, the entirety of the observable universe is liquid in nature, and—moreover—replete with color.

Hypothetically, of a being who is able to see the entirety of the observable universe that comprises the being's visual-field, the being's visual-field would be replete with a conglomeration of integrated and interacting objects such that the being would not be able to see beyond what is immediately in front of the being's eyes, perhaps tantamount to what a person of the observable universe would see upon being placed into a pool of integrated and interacting opaque liquids of an immense array of colors: The person would only be able to see the various liquids that are adjacent to his or her eyes. And for the purpose of this analogy, the person, while he or she is observing what occurs, would not encounter any degradation of the typical functioning of his or her eyes and vision.

56. That it can be stated with mathematical language that a particular point, that is located at a particular place, has zero depth, zero length, and zero height, and that a particular line, that is located at a particular place, has X length, zero depth, and zero height, and that a particular, geometric shape, that is located at a particular place, has X length, Y height, and zero depth, does not necessarily mean that there are such things, or that there could be such things, and that such things can be accurately pictorially represented, and coherently pictorially conceived.

Devising the antonymic mathematical-linguistic alteration of a zero-dimensional existent, and the subtractive mathematical-linguistic alterations of one-dimensional and two-dimensional existents, for the purpose of concluding that what was uncovered during various kinds of experimentation are fundamental constituents of the universe, is both self-deceptive and publicly deceptive.

Devising ingeniously furtive, dissociative, incoherent, and negligent theory in order to conclude that one's experimentation has succeeded in accomplishing the primary goal for which it was designed is seemingly dishonest. Moreover, devising such theory for the purpose of obviating, or substantially forestalling, experimental confirmation is also seemingly dishonest.

57. In light of *(55),* and *(37)*, it the case that the observable universe is additionally replete with constituents; and of course the microlimits of the constituents is space, and that the observable universe is also replete with space.

It is therefore case that there are no true macrolimits of the constituents of the observable universe, as there is of the observable universe itself, but rather, only relative macrolimits that are, moreover, completely integrated with other constituents of the observable universe, such that there is no true division between any of the constituents. For there to be true divisions between the constituents of the observable universe, there would have to be external universe between them; and they would, as such, be discrete observable universes themselves.

The previously discussed issue of whether the macrolimits of the constituents of the observable universe could be observed from the other side of the microlimit of the observable universe, or the other side of the macrolimit of the observable universe, or perhaps both, is therefore not an issue, as the constituents of the observable universe cannot have true macrolimits, but rather, only relative macrolimits.

58. The microlimit of the constituents of the observable universe—that is, space—is what integrates the constituents.

59. It is surely the case that the light-spectrum (the electromagnetic spectrum) exceeds what is currently measured to be the shortest wavelength of light, and what is currently measured to be the longest wavelength of light; and in order to know whether the current shortest wavelength of light, and the current longest wavelength of light, and the speculated shortest wavelength of light, and the speculated longest wavelength of light, are the shortest and longest of the observable universe, it would necessary to exceed the observable universe, and, moreover, become formally different in nature. It is moreover surely the case that the light-spectrum consists of additional general kinds of light than what is currently measured to be the general kinds of light, and additional general kinds of light beyond what are currently generally or precisely measured to be the microlimit and macrolimit of the light-spectrum: Within the visible light-spectrum, for example, there surely is an extraordinary extent of smaller (or narrower) wavelengths of light that comprise a multitude of different general kinds of light; and as an aside, such light would be a profusion of colors that humans have never seen, nor can, as such, imagine.

In order to measure, and as such know, the complete intricacy of the light-spectrum, it surely would be necessary to exceed the observable universe, and, moreover, become formally different in nature: Beings of the observable universe are formally constrained from precisely sensory-perceiving, measuring, and conceiving of the fundamental features of the observable universe, including their minds; and it is likely the case that from the perspective of what exceeds the observable universe, and in particular, of beings who are formally different than the beings of the observable universe, the fundamental features of the observable universe could be precisely sensory-perceived (or experienced in another way), measured, and conceived of.

The microlimit wavelength of the light-spectrum, and the microlimit size (or narrowness) of each wavelength of the light-spectrum, is surely the microlimit of the observable universe.

The macrolimit wavelength of the light-spectrum is likely akin to the macrolimit of any constituent of the observable universe.

Hypothetically, of any wavelengths of the light-spectrum that reach the macrolimit of the observable universe, or which would do so, their macrolimits would be the macrolimit of the observable universe.

What light precisely is surely could only be seen and, or, measured from the perspective of beings of the external universe. If we could see and, or, measure light itself, we surely would not be able to see and, or, measure what we do of the observable universe.

60. Non-visible light—that is, light that does not, for humans, illuminate the constituents of the observable universe—surely does illuminate the relative space of the observable universe, and the constituents of the observable universe, and with the expression of a profusion of colors that humans cannot see, nor imagine.

Hypothetically, for sighted beings who do not see the expression of the region of the light-spectrum that we do, and who, instead, see the expression of a different region of the light-spectrum, they would surely see the constituents of the observable universe, and, at times, the relative space of the observable universe, as being of a profusion of colors; and the colors would surely be ones that we cannot see, nor imagine.

The illumination of relative space with the expression of the colors of light is discussed in *(62)*.

It is of course possible that some or all of the other sighted beings of the observable universe, and some or all of the sighted beings of the external universe—if there are such beings—do see, or would see, the colors that we see of our visible-light as being different than any of the colors that we see; and the colors would of course be ones that we cannot see, nor imagine. It is also possible, of course, that our visible-light is, to some of the sighted beings of the observable universe, and to some or all of the sighted beings of the external universe—if there are such beings—non-visible light (that is, not illuminating of any features of the observable universe).

61. Since, as discussed previously, space itself cannot have internal curvature, nor any other internal geometric form, and can only have curvature, or other, geometric forms, at its macrolimit, gravity is a phenomenon of the condensation of space. And as such, of regions of space in which gravity is comparatively stronger, space is comparatively more condensed. And of regions of space in which gravity is comparatively weaker, space is comparatively less condensed. Perhaps analogously, a comparatively more condensed liquid would entail stronger gravity within the liquid, and a comparatively less condensed liquid would entail weaker gravity within the liquid.

The condensation of space is therefore what causes gravitational lensing. Moreover, the concept of condensed space provides the most coherent explanation of gravity and gravitational phenomena.

The condensation of space would be, like space itself, something that cannot be accessed in order to determine what it is.

62. The observed blackness of any region of relative space is due to the presence of light, for otherwise there would be blindness instead of blackness. The black of the visual-field is therefore produced by the expression of one or more light-sources. (The black of the visual-field is of course not revealed by common "visible-light" (that is, light of 400-700nm, which is actually non-visible light, as it itself cannot be seen): Only the black constituents of the observable universe are revealed by such light).

The black of the visual-field is therefore light itself, and could be referred to as "black light."

Light pervades the entirety of the observable universe, by virtue of how it is the observable universe.

In a room whose enclosure provides the most substantial visible-light light-blocking, and non-visible-light light-blocking, that are available, the darkness that would be seen may be caused by black light that is expressed from one or more sources that are outside of the enclosure.

If one became blind upon entering a light-blocking enclosure, then the enclosure succeeded at either blocking the entirety of light, or a considerably greater extent of it than can be accomplished anywhere else.

Relatively pure darkness is not obtained by the relative absence of light, but rather, by the relative absence of human visible-light: The black that is seen is one or more other kinds of light.

Unlike our visible-light, which itself is invisible to us, but whose sources are visible to us, black light is visible to us; and in light of how black light is visible to us, the sources of black light are likely invisible to us.

The black of the black constituents of the observable universe is surely not illuminated by visible-light, due to how visible-light consists of an array of colors that does not include black. The black of the black constituents of the observable universe is therefore always illuminated as black; and the constituents of the observable universe that are black are seen as black only when the observable constituents that surround them, and which comprise their backgrounds, are of disparate colors, and are illuminated by 400m-700nm light.

It is surely the case that black holes are not objectively black; and it is likewise surely the case that there are no objective particular colors, but rather, only that all constituents of the observable universe are colored.

It is surely the case that there are no objective visible-light sources, but rather, only light sources that are visible-light sources to particular kinds of beings.

Given the pervasiveness, uniformity, and persistence of black light throughout the predominance of the observable universe, it may be the case that the source or sources of it are in the external universe: If the source or sources of it were in the observable universe, surely it would, due to it, and its sources, being subjected to phenomena of the observable universe, be at least intermittently less pervasive, uniform, and persistent in various areas of the observable universe.

Black light surely pervades the entirety of the observable universe; and if it does, portions of it would therefore be supplanted by visible-light of at least a particular minimal intensity, as well as by one or more particular regions of visible-light of at least a particular minimal intensity, as well as by at least a minimal illumination of the observable constituents of the observable universe whose surfaces are not black. The extent of the supplanting is, of course, partly dependent on the nature of people's vision; but in general, particular intensities of the aforementioned kinds of light succeed, for the predominance of humans, at, for example, (a) minimally supplanting portions of black light to the extent that at least a semblance of most of the observable constituents of the portions of the visual-field, and the relative space between the constituents, can be seen, (b) essentially completely supplanting portions of black light to the extent that all of the observable constituents of the portions of the visual-field, and the relative space between the constituents, can be seen, and (c) not only completely supplanting portions of black light, but overriding the portions of the visual-field to the extent that the portions of the visual-field, including the surfaces of all of the observable constituents of the portions of the visual-field, appear, by way of reflection (for the constituents), and by way of illumination (for the portions of relative space between the constituents), as light-sources themselves.

Hypothetically, it would surely be the case that upon the removal of all of the content of the observable universe, except the microlimit and macrolimit, the supplanting of black light would be discontinued, and black light would, as such, be completely pervasive; and this indicates that the source or sources of black light are in the external universe.

63. The feature of the observable universe that is either the black macrolimit of the observable universe, or which in part consists of it, may be the source of the black light of the observable universe.

64. Instead of, "The black of the visual-field is therefore light itself, and could be referred to as "black light"," the black of the visual-field may be the illumination of the microlimit of the observable universe (that is, the illumination of space itself); and the source of the illumination may be the macrolimit of the observable universe, or an aspect of it. Since the microlimit of the observable universe must be physical in nature, it is surely the case that the black-illumination of the relative space of the observable universe is the illumination of the constituent, or constituents, that comprise the microlimit.

It is of course possible that black light, the black microlimit, and the black macrolimit, are not objectively black, but rather, black only to us, and to perhaps most, or all, sighted beings of the observable universe.

It is also possible that the source or sources of black light are located at the other side of the microlimit of the observable universe.

It is also possible that the source of black light is the microlimit itself.

It cannot be the case that space itself is black apart from any illumination of it: Hypothetically, if the content of the observable universe was removed, except of course the microlimit and macrolimit, and if, as such of course, there was a discontinuation of any sources of visible-light, and if there were no sources of light in the external universe, and if the microlimit and macrolimit were not sources of light, and if the microlimit and macrolimit did not in part consist of one or more sources of light, there would, for visual beings, be blindness, rather than darkness.

65. An addition to *(22)*: In order to observe and, or, measure whether the form of observable universe (the microlimit and the macrolimit) is expanding, or diminishing, and likewise, in order to observe and, or, measure whether the observable universe itself is moving, it would be necessary to exceed the observable universe.

While it can be coherently speculated that the content of the observable universe, or, a portion of the content of the observable universe, may have been of a general size, at a general time, in the past, and may be of a general size currently, and may be of a general size, at a general time, in the future, this is irrelevant to what the general size of the form of the observable universe may have been at that time in the past, and may be currently, and may be at a general time in the future.

It cannot be coherently speculated that the form of the observable universe was of a general size, at a general time, in the past, and is of a general size currently, and will be of a general size, at a general time, in the future.

In light of what was discussed previously about "infinity," "eternal," etc., the observable universe (both its content, and form) surely did have an origin; and there may have been different origins for each—that is, the form of the observable universe may have originated first, rather than both concurrently.

66. Of anything of the observable universe that can be *precisely* observed and, or, measured, or inferred to exist based on precise observations and, or, measurements of other features or phenomena of the observable universe, it is not a fundamental feature of the observable universe: Anything that can be precisely observed and, or, measured could possibly be physically investigated via penetration, disintegration, etc., thereby revealing that it is not of a fundamental nature; and anything that is inferred to exist based on precise observations and, or, measurements of other features or phenomena of the observable universe would be of the same formal nature, though perhaps comparatively more simple, as the features or phenomena that the inference is based on.

67. The external universe, in necessarily being formally different than the observable universe, surely does not have the causation that is the case of the observable universe: For example, as to how the microlimit of the observable universe is a three-dimensional boundary, rather than a boundary-edge of a three-dimensional realm, this formally not only cannot be physically nor technologically investigated, it cannot be conceived of.

68. What are referred to as "black holes" are likely a cause, or the cause, of black light.

69. Black light is visible-light.

Black light can be supplanted by a sufficient extent of light that is of what is commonly referred to as the "visible-light spectrum."

70. The microlimit and macrolimit of the observable universe, while of course physical in nature, are not *of* the observable universe, in the sense of (a) not being the same in general nature as the features of the observable universe, namely, conceivable, analyzable, potentially analyzable, precisely observable, potentially precisely observable, etc., and (b) not being caused by, sustained by, nor affected by, the features of the observable universe.

Again, there are no macrolimits of the constituents of the observable universe.

The microlimit and macrolimit of the observable universe, while of course physical in nature, must be formally different than any feature of the observable universe.

71. An experiment that may in part demonstrate that one region of space is more condensed than another region of space is that of subjecting the different regions of space to the same extent of the aforementioned visible-light illumination, and assessing whether there is a difference in the extent that the regions of space and, or, the constituents of the regions of space, are illuminated.

I speculate that black holes predominately or entirely consist of comparatively highly condensed space, and that they, as such, have comparatively extremely strong gravity.

I speculate that due to the highly condensed space of black holes, it hypothetically would be relatively easy to illuminate regions of black holes with the aforementioned visible-light. However, black holes are completely and intensely illuminated with black light; and the black light surely supplants the aforementioned visible-light, analogous (but opposite) to how, on earth, minimal amounts of visible-light supplant black light. And again, black holes are likely the sources of their own illumination.

Visible-light would travel faster in regions of less condensed space; and as such, it would illuminate those regions comparatively less. However, since the black light of comparatively less condensed space is less intense, the black light is easily supplanted by visible-light.

72. As for what of black holes would produce black light: Since it is likely the case that there are no features of black holes except for (a) that they are composed of space, and, as such, replete with the microlimit of the observable universe, and (b) that they consist of comparatively highly condensed space, it is likely the case that the source of the black light is from (1) what exceeds the microlimit of the black holes, or (2) the distinctive space (that is, the distinctive microlimit) of the black holes. Regarding (2): Whereas, as was discussed earlier, all of space (the microlimit) expresses black light, and in so doing, illuminates itself, the regions of space that are comparatively highly condensed may express black light of greater intensity than regions of space that are comparatively less condensed.

Black holes could be conceived of as black stars.

Relatedly: The entirety of the observable universe must be one physical unit, as, again, there are no macrolimits of any of the relative constituents and phenomena of the observable universe. The entirety of space, therefore, is likely of one black star; and different regions of space, due to being condensed differently, would express different extents of black light.

The entirety of the observable universe is, therefore, likely one black star (or one black microlimit star); and the relative constituents and phenomena of the observable universe are likely constituents of the star.

73. Another experiment that may in part demonstrate that one region of space is more condensed than another region of space may be that of subjecting the different regions of space to the same extent of the aforementioned visible-light illumination, and assessing whether there is a difference in the speed that the light travels through the regions of space. Where space is comparatively less condensed, and where, as such, gravity is comparatively weaker, light would travel faster than it would in regions of space that are comparatively more condensed, and where, as such, gravity is comparatively stronger.

74. What is referred to as "dark matter" is surely the microlimit, and different condensations of the microlimit; and what is referred to as "dark energy" is surely black light, and different intensities of black light.

75. While common visible-light can supplant black light, black light is still present, as is demonstrated by how a container that blocks 100% of common visible-light, and which is surrounded by profoundly intense common visible -light, consists of 100% black light inside. If common visible-light completely supplanted the transmission of black light across the visual field, then for someone inside the aforementioned container, there would be complete blindness (which, as discussed before, does not entail black-vision).

It may be the case, though, as was discussed, that space is a, or the, source of black light.

76. Black light is visible-light (that is, it itself can be seen), whereas common visible-light is invisible (that it, it itself cannot be seen).

Black light is a facet of the light spectrum.

Since black light is observable, if its sources are black holes, it surely could be measured, since the sources would be of the observable universe. If black light is only a feature of the microlimit of the observable universe, and, or, the macrolimit of the observable universe, and, or, the external universe, then it surely could not be measured, because these aspects of the universe are not *of* the observable universe, as was discussed.

There are things in the observable universe that can be imprecisely observed, but which cannot be measured, because they are not of the observable universe: For example, space, and the macrolimit of the observable universe, via black light illumination; and perhaps black light itself. The condensation of space, itself, cannot be observed; and any measurements of it would consist of ascribing measurements to regions of space based on what occurs to various existents in those regions of space relative to what occurs to those existents in other regions of space. For example, if light slows to X degree, and, or, curves to X degree, in a particular region of space, then the region of space could be ascribed a measurement that represents this. However, the condensation of space, itself, cannot be measured.

77. Objects of the observable universe that spin on their own accords, and which orbit on their own accords, do so because they are spun, and orbited (put into orbit), by the perpetual spinning of space around, and within, them, and the perpetual circulation of space in their orbits, and within them.

The earth orbits because of the presence of space that circulates in that orbit: As discussed earlier, space is within all of the constituents of the observable universe; and the space that circulates in the earth's orbit includes the space of the earth.

The process or processes that cause space to circulate are of space itself (that is, the microlimit itself), and, or, what exceeds the micro-limit, and not of the observable universe.

Likewise, the earth spins because it is spun by space.

The aforementioned explanations are superior to the prevailing explanations that the earth spins and rotates as it does due to inertia.

78. Where I previously discussed "black light illumination," it is perhaps more accurate to describe the phenomenon as "black light expression": The microlimit and macrolimit likely express black light, thereby revealing that they exist, and are black. (Again, if there was no macrolimit, the background of the observable universe would be a region of blindness; and if there was no microlimit, our visual-fields would be regions of blindness. Aside, I previously argued, regarding the hypothetical absence of the microlimit, that there would be blindness between objects of our visual-fields; but I realize that there would also be blindness between our eyes and objects).

79. It is possible that the macrolimit is a region of immensely condensed space (that is, immensely condensed microlimit), and, moreover, and as such, that the light from the objects that are beyond our solar system is subjected to such extreme gravitational lensing that the objects are not even approximately where they seem to be, and not even approximately at the distances that they seem to be.

80. An addition to *(36)*, *(43)*, and *(44)*: "Infinity" (the term and symbol) actually mean the following:

(1) Finite processes that are conceived of to proceed indefinitely despite how, at any moment, their procession can be observed and measured, or, in the absence of observation and measurement, conceived of, to be at a particular level or degree of procession.

(2) Finite states (or realms) that are conceived of *to have already proceeded indefinitely (or to infinity)*, despite (a) that processes, at any moment, are finite, and (b) that the concluding states of any processes, by virtue of being states of processes, can be observed, measured, or conceived of to be finite.

(3) Finite states (or realms) that are expressed as, "Realms that have no outer boundaries," where the *belief* in "no outer boundaries" is (a) a severe dissociation from the general concept of a realm, namely the dissociation from how realms, by virtue of being realms, have general boundaries, and (b) conceptually meaningless, because nothing conceptual is expressed by "no outer boundaries"—that is, the phrase is a meaningless antonymic alteration; and it is, as such, only psychologically (emotionally) meaningful, in the sense that it expresses emotion, and elicits emotion. (Aside: Language, including mathematical language, can be altered in ways that result in a partial or complete loss of conceptual meaning, but a concurrent eliciting of immense emotion; and in this case, the linguistic alteration, "A realm that has no outer boundaries," may elicit a combination of awe, incredulity, wonder, excitement, etc).

81. Notwithstanding that the microlimit and macrolimit of the observable universe are physical in nature, their physicality is formally different than anything of the observable universe, due to how they, while they are *in* the observable universe, are not *of* the observable universe. As such, it is likely impossible to measure any facet of the microlimit and macrolimit, such as their black light expression, and their physicality.

Since black light is likely not of the observable universe, but rather is expressed by the microlimit and macrolimit, or, what exceeds one of them, or what exceeds both of them, it is formally different than any light of the observable universe. As such, it is likely impossible to measure black light; and a practical indication of this is that, during optical experimentation that entails blocking immense extents of common visible-light and common non-visible light, or essentially the entirety of both, a state of blindness never occurs, nor partial blindness (for example, when one's visual-field consists of a dispersion of minute regions of sight and minute regions of blindness).

Again, if the microlimit and macrolimit, black light, and condensed space, could be precisely observed (that is, observed for what they actually are), and precisely measured, this would mean that they are not fundamental features of the observable universe, and that they are, instead, both *in* the observable universe, and *of* the observable universe. Again, for example, the observable universe's microlimit and macrolimit cannot be of the observable universe in order to be limits of the observable universe: Anything of the observable universe can possibly be disintegrated, divided, penetrated, etc., to reveal comparatively more fundamental features; and anything of the observable universe can, moreover, be conceived of as being disintegrable, divisible, penetrable, etc.

82. Notwithstanding that common visible-light (which, again, is itself invisible) can partly or completely supplant black light for human vision, I predict that black light remains expressed throughout the entirety of the observable universe, and that this could be observed by way of extreme microscopy: At a particular level of microscopy, black light should be observed regardless of the extent of visible-light.

When extreme microscopy (or super-resolution microscopy) gradually reaches complete blackness even in the presence of extreme common visible-light, this likely means that the microlimit can be more closely observed. However, it (the microlimit), notwithstanding, could only be observed as black, as can be done now by human vision in the absence of common visible-light.

83. Professor Lisa Randall stated the following about dark matter:

> We know that there is dark matter. Stuff that is really matter. Clumps, etc. But, it doesn't emit light: We don't know what it is. It's probably not one of these Standard Model ingredients. In fact, it's definitely not one of these Standard Model ingredients.[3]

As was discussed, it, or the external universe that exceeds it, or both, emit black light.

> All we know about it is that it's matter. It doesn't interact with light; at least at the level we've studied it. And it interacts only a little bit with itself.[4]

> Why we call it dark matter is that it doesn't interact with light, as far as we know. So in other words, light just passes through it. We should have called it transparent matter, because dark stuff absorbs light; but this stuff, light just passes through.[5]

---

3. Lisa Randall on, "How is the Cosmos Constructed?" 2:14-2:29, *Closer To Truth*, www.closertotruth.com/interviews/2876, 2017.
4. CajaCanarias Foundation, Interview with Lisa Randall, Video: *Turning On the Cosmos,* 4:40-4:52, https://youtu.be/juZ5nfvhhCU, 2015.
5. Charlie Rose, Video: Interview with Lisa Randall, 1:10-1:28, www.charlierose.com/videos/23185, 2015.

The black light that it emits interacts with common visible-light, in the sense that common visible-light can partly to completely supplant it for visual beings. However, as was discussed earlier, black light (and, as such, of course, the microlimit itself) are persistently present throughout the entirety of the observable universe. (The microlimit (space) is dark matter; and the black light that it emits is dark energy).

84. The true color of all of the relatively separate existents in the observable universe is black: The microlimit is pervasive throughout the observable universe, including in all of its relatively separate existents; and as such, black light is expressed by all of the relatively separate existents of the observable universe. Humans, and all other existents, including stars, and other true non-visible light sources, are pure black. Extreme microscopy would reveal that even existents of relatively extreme non-visible light are fundamentally purely black.

85. For existents that remain relatively black in the presence of common intensities of common visible-light (hereafter I will refer to "common visible light" as "true non-visible light"), it is necessary to subject such existents to greater intensities of true non-visible light in order to partly supplant their expression of black light. Subjecting the black existents to particular extents of true non-visible light will, for human vision, result in the surfaces of the existents appearing as reflections of the true non-visible light. The existents' expression of black light is never fully supplanted, though, as extreme microscopy would reveal that the existents are inexorably pure black.

86. Existents whose surfaces appear to be of non-black colors are so because the surfaces are of matter that absorb particular kinds of true non-visible light, and reflect other kinds of true non-visible light. (Of course, though, a particular intensity of true non-visible light, and all intensities greater than it, would result in the existents' surfaces appearing as reflections of the true non-visible light). The existents, though, do not *express* the non-black light, but rather, reflect true non-visible light: It is the reflection that entails that their surfaces appear, for visual beings, of particular non-black colors. Regardless of the reflection, though, the existents are fundamentally black, and always, as such, express black light. And in the approximate absence of, or essentially complete absence of, true non-visible light, the existents of course are their true colors, namely black, which of course is caused by their expression of black light.

87. When super-resolution microscopy gradually reaches complete blackness even in the presence of extreme common visible-light, the microlimit can be more closely observed. However, it (the microlimit), notwithstanding, could only be observed as black, as can be done now by unaided vision and aided vision in the absence of common visible-light.

When particle division experimentation, and, or, particle collision experimentation, and similar experimentation, encounter circumstances in which (a) what is attempted to be divided, collided, etc., seemingly vanish, and what remains is black light, or (b) aspects of what is divided, collided, etc., seemingly vanish, and what remains is black light, the experimentation has succeeded at reaching the microlimit of the observable universe more closely than what can be done by way of unaided vision and aided vision. However, what is reached is the same in form as what is seen by way of unaided vision and aided vision.

What is the ideal outcome that super-resolution microscopy and extreme telescopy could accomplish?: Of any relative existent that is observed via microscopy, it, by virtue of being an existent, has an intricacy which, moreover, could be more closely observed by more powerful microscopy; and without having observed the macrolimit of the observable universe via telescopy, it cannot be concluded that any object or objects observed by telescopy are located at the macrolimit. The ideal outcome of super-resolution microscopy and telescopy is to arrive at pure black light. Such light is an aspect of the microlimit and macrolimit of the observable universe.

88. Professor Leonard Susskind states the following about "extra dimensions":

> The extra dimensions of string-theory must be rolled up and curled up into some tiny little dimensions that we can't see because they're too small. …. Let's suppose that the world, at it's smallest distances, has more than three dimensions; and these extra dimensions are curled up in little knots—pretzels, doughnuts, whatever we want to think about them …. and only the tiny creatures are small enough to move around and see them, creatures by which I mean elementary particles.[6]

However, he then states the following:

> They *could* be big. They could be small. [He then states the following in a sarcastic tone-of-voice in order to criticize the supposition that extra dimensions must be small]: "I can only find three of them! I can't seem to find the other one's! They must be very small!" The way to think about it is: Imagine a one-dimensional world—imagine a world on a line; we could have particles, maybe beads, moving up and down that line; but then some ultra-small creatures discover that they don't live on a line—they live on a cylinder; they can move around the cylinder as well as along the cylinder; they have discovered that their world is really two-dimensional, not one-dimensional. They have discovered an extra dimension. But, the big clumsy creatures that might be made out of these beads, they can't see

6. Leonard Susskind on, "How Do Particles Explain the Cosmos?" 1:16-2:10, *Closer To Truth,* www.closertotruth.com/interviews/3078, 2016.

> the extra small dimensions. Only the little tiny guys can see hem.[7]

As such, he reverts to his initial argument that extra dimensions "must" be small.

The aforementioned Professor Lisa Randall similarly states the following about "extra dimensions":

> One of the best ways to imagine how we should envision extra dimensions was done by Edwin Abbot Abbot—late nineteenth century in a book called Flatland. And what he said is: If you lived in two dimensions, you would have the same difficulties picturing and imagining the third dimension that we have imagining a fourth spatial dimension, or higher dimensions. So what he said is: Suppose that there were creatures living on Flatland, say this tabletop; and how would they see the world. I mean, we know that there is a third dimension; but these Flatlanders wouldn't be aware of it. So what would they see? Well suppose that some object, like from the third-dimension came through. Suppose that a sphere came through. A Flatlander would never see the sphere in its entirety. They can't see three dimensions, the same way that we can't see more than three dimensions. But what they would see is a series of disks that grew in size and then decreased in size as the sphere went through. So they would get all of the information about this three-dimensional object, even though they couldn't necessarily reconstruct it in their pictures to see that it was a three-dimensional object. So in the same way it could be that there are extra dimensions that we don't see directly.[8]

---

7. Leonard Susskind on, "How Do Particles Explain the Cosmos?" 5:05-5:52, *Closer To Truth,* www.closertotruth.com/interviews/3078, 2016.

8. Lisa Randall on, "Are There Extra Dimensions?" 0:38-1:44, *Closer To Truth,* www.closertotruth.com/interviews/2874, 2017.

However:

(1) Spatial dimensions, themselves, (a) are not existents, and (b), and as such, are not the kind of thing that can be "rolled up," "curled up," etc: Spatial dimensions are attributes of existents, and, as such, are dependent for their existence (their existence as attributes) on the existents; moreover, only existents can be rolled up, curled up, etc. Conceiving of spatial dimensions as existents is to dissociate the dimensional attribute that existents have from the existents. This is analogous to conceiving of existents' solidity, color, weight, temperature, etc., as existing independently from the existents. Is solidity an existent? Can color exist independently of any existent? Expressed more generally: Can attributes of existents exist independently of existents? Attributes, by definition, are qualities of existents; and as such, attributes depend entirely on the existence of the existents for their existence: Without the existents, they do not exist. Aside: Space, as I discussed earlier, is an existent; and as such, dimensions do not exist independently in contexts in which there are no common existents—that is, common tangible existents, common visible existents, etc.

(2) Notwithstanding that Susskind may not be arguing the following, the curling up, rolling up, etc., of existents cannot entail that they have extra spatial dimensions; and stating that the extra spatial dimensions are hidden within the curled up existents, and cannot as such be observed, is deceptive. (a) Regardless of how existents are curled up, rolled up, etc., they are three-dimensional: There is no special way of curling up existents that would entail that they have extra spatial dimensions. (b) No pictorial representation of existents as having extra spatial dimensions has ever been made, nor can be made. (c) As discussed in *(16)*, *(17)*, and *(18)*, mathematical language that is argued to express extra spatial dimensions is actually language of additive linguistic alterations, which, while emotionally puissant, are conceptually meaningless, pictorially meaningless, and linguistically meaningless; and again, not all mathematical language, simply by virtue of it being

mathematical language, expresses something: A considerable extent of mathematical language, and the ensuing mathematics, is entirely psychological, or psychological to different degrees.

(3) I will also refer to my previous discussions about extra spatial dimensions, and extra non-spatial dimensions, namely my discussions at *(16)*, *(17)*, and *(18)*.

(4) Regarding Randall's invoking of the 1884 novel, *Flatland: A Romance of Many Dimensions*, as an analogy, and Susskind's similar analogy, both Randall and Susskind conflate people's spatial circumstances and their neurological circumstances: Both assume that if entities (entities who, at a minimum, are capable of vision and movement) reside in a "one-dimensional" or "two-dimensional" world, they would not be capable of "seeing" the third dimension (whether that be three-dimensional objects that enter their worlds, or the three-dimensional world that exists *with* their worlds. However, (a) the nature of people's spatial circumstances does not entail that their neurologies (their sensory abilities) are such that they can see only X-dimensional objects, or, an X-dimensional world, (b) the nature of people's spatial circumstances does not entail that their minds (their conceptualization and imagination) are such that they can only conceive of, and only imagine, X-dimensional objects, or, an X-dimensional world, (c) the beings that Randall and Susskind discuss have the capacity to see and move, and are often intimated to be human or human-like; and as such, in addition to having the capacity to conceive of, and imagine, three-dimensional objects, they surely have the capacity to pictorially represent three-dimensional objects via drawing, and (d) Randall and Susskind assume that beings who live in an X-dimensional world also have X-dimensional vision; and in addition to this being false (as discussed above), there are no such things as 1-dimensional and 2-dimensional vision, nor can there be.

(5) Randall and Susskind simultaneously state that the beings of one-dimensional or two-dimensional worlds (a) "live" in such "worlds,"

and (b) the beings of such worlds (for Randall) cannot see three-dimensional objects, as three-dimensional objects, that enter their worlds, and, for both Randall and Susskind, they cannot see the three-dimensional world that exists *with* their one-dimensional or two-dimensional worlds. (a) For beings who live in one-dimensional or two-dimensional worlds, there is no three-dimensional world, nor three-dimensional objects, available to see. Moreover, it would not be possible for a three-dimensional object to enter into such a world. Again Randall states, "Well suppose that some object, like from the third-dimension came through. Suppose that a sphere came through." (b) Randall and Susskind should have stated that the beings live in three-dimensional worlds, and can only see the one-dimensional or two-dimensional aspects of the three-dimensional worlds. (c) Since the beings neurologies are not affected by their spatial circumstances, while they themselves are spatially one-dimensional or two-dimensional, and while, as such, their ability to *move* is limited by their spatial circumstances, their neurologies (their sensory abilities), and their minds (their conceptualization and imagination), would allow them to see, conceive of, imagine, and draw, the three-dimensional. (d) The three-dimensional can be readily conceived, imagined, and pictorially represented, even by beings who are spatially limited by their spatial circumstances.

(6) Spatial dimensions, unlike existents, (a) cannot be small or big, and (b) cannot exist in some size-scales, and not in other size-scales: Spatial dimensions, themselves, do not have size, as, again, they are attributes of existents; and only existents can have size. Moreover, all of the spatial dimensions of a realm, such as the observable universe, are attributes of that realm that are pervasive throughout that realm.

(7) If hypothetically there were extra dimensions at a minute scale, and if the observable universe was, as such, fundamentally composed of existents (e.g., the strings of String Theory) that are of extra dimensions, then all of the constituents of the observable universe would be fundamentally composed of existents of extra dimensions. This would,

as such, certainly affect the physical mechanics of the constituents: While it would seem to us that the observable universe is three-dimensional, we and all other existents would not be able to mechanically interact with the observable universe in a way that is explained by it being three-dimensional: There would be immense, mechanical disruption that could not be explained, and which interfered greatly with the mechanics of all of the constituents. Moreover, I would speculate that we would not exist if either, or both, we and the rest of the observable universe was fundamentally composed of extra-dimensional existents.

(8) Actual spatial dimensions can be geometrically represented via pictorializations, and clearly conceived of, and imagined.

(9) Susskind and Randall discuss spatial dimensions as if the observable universe consists of combined worlds of spatial dimensions, namely the one-dimensional world of spatial dimension, the two-dimensional world of spatial dimension, and the three-dimensional world of spatial dimension, and possibly the zero-dimensional world of spatial dimension. As discussed in *(16)*, *(17)*, and *(18)*, separating the three spatial dimensions from one another is to engage in severe dissociation; and there cannot be zero-dimensional, one-dimensional, two-dimensional, four-dimensional, etc., worlds, nor can such worlds be coherently conceived of. Moreover, expressions of zero-dimensionality are antonymic, linguistic alterations which, as such, are conceptually, pictorially, and linguistically meaningless.

It is therefore the case that the analogies that are constructed by Randall and Susskind—and those analogies, and other analogies that are the same in form, are used by a multitude of physicists and others—do not succeed at demonstrating that it is possible that there are extra dimensions in the observable universe. Moreover, the analogies do not succeed at demonstrating that it is conceivable that there are extra dimensions: I discuss this in *(16)*, *(17)*, and *(18)*.

89. Professor Michael W. Davidson, et al., state the following about microscopy; and this supports my discussion in *(87)*.

> The brightness or radiance of an illumination source designed for use in optical microscopy is one of the most important characteristics to be considered due to the fact that the intensity of an image is inversely proportional to the square of the magnification according to the equation:
>
> $$\text{Image Brightness } \mu \ (NA/M)^2$$
>
> where NA is the objective numerical aperture (in effect, the objective's light-gathering ability) and M is the magnification. Thus, as the objective magnification is increased, image brightness is proportionally decreased depending upon the numerical aperture.[9]

And,

> .... image brightness decreases rapidly as the magnification increases ....[10]

The above general principle can be extended to predict that at a particular level of super-resolution, no light source (that is, illumination source) will succeed at illuminating what is being attempted to be observed. The level of resolution is such that all of the true non-visible light (that is, common visible-light) is, so to speak, looked beyond. That is, the resolution exceeds the realm of the true non-visible light. That is, the true non-visible light is capable of extending only so far into the visual-field; and the level of resolution is capable of extending beyond that distance.

---

9. Michael W. Davidson, Andreas Nolte, and Lutz Höring, "Fundamentals of Illumination Sources for Optical Microscopy," Zeiss, http://zeiss-campus.magnet.fsu.edu/articles/lightsources/lightsourcefundamentals.html.

10. Michael W. Davidson and Kenneth R. Spring, "Image Brightness," MicroscopyU, www.microscopyu.com/microscopy-basics/image-brightness.

In the above circumstance, what is observed is pure black light. It is not the case, though, that the light source is simply incapable of illuminating what is being attempted to be observed. Rather, the light source should succeed at accomplishing the requisite illumination, but it does not; and this is because the level of resolution has succeeded at looking beyond all of the local micro-existents of the observable universe, and into what remains, namely the black light that is emitted from dark matter (which hereafter I will refer to as “black matter”).

At a particular level of microscopy, and at all levels thereafter, only black light should be observed. Black matter, of course, is present, and pervasive throughout the entirety of the observable universe, and is what emits black light; but because it is formally different in nature—that is, because it is of a formally different kind of matter than any of the matter *of* the observable universe, and, as such, of a different kind of physics—it could never be observed. And again, while it is *in* the observable universe, it is not *of* the observable universe.

The following hypothetical is one in which what is observed via super-resolution microscopy is its own light source, namely a fragment of a star. While of course the experiment is hypothetical, I predict that at a particular level of resolution, the extreme light of the star would be exceeded, and all that would be observed is pure black light. Stars, like all of the other relatively separate existents of the observable universe, are fundamentally pure black; and it is only the presence of 400nm-700nm light that, for visual beings, causes their surfaces, via partial reflection and partial absorption, to, at the time of the presence of the light, be of different colors.

90. Principles of optical geometric physics:

(1) All geometric forms, from the simplest, namely a point, to the complex, are three dimensional, as was discussed.

(2) The macrolimits of geometric forms are macrolimits by virtue of the presence of relative space, and, or, relative objects, that are

outside of the macrolimits. Hypothetically, if there was nothing that is outside of what are now found to be macrolimits, (a) the only perspective that would remain is the perspective from within the geometric forms; and from that perspective, it could not be observed, measured, nor accurately conceived of that the macrolimits that are observed, measured, and conceived of are, in fact, the macrolimits, and (b) what may be observed, measured, and conceived of to be the macrolimits would in fact not be the macrolimits, but rather, the limits, or ends, of spatial reality.

(3) From the perspective of within a geometric form whose interior is devoid of light sources and objects—and therefore devoid of non-black color—(a) there must be a microlimit in order for there to be an interior; and the only conceivable microlimit is one of (a^1) the intangible, unobservable, unmeasurable, and inconceivable black matter, and (a^2) the partly observable and partly conceivable, but intangible and unmeasurable, black light, both of which would of course pervade the entirety of the interior, (b) without black matter, there would not be an interior, and (c) without black light, there would be blindness, despite having the ability to see.

(4) From the perspective of within a geometric form whose interior consists of extreme-intensity light sources that are near to, and far from, an observer, if hypothetically a portion of the macrolimit was removed, and if only the far light sources were emitting light, and if there was external universe on the other side of the macrolimit, then when that region of the visual-field was looked at, it would be experienced as a region of blindness: While the interior of the geometric form is saturated with black matter and black light, (a) the black light is easily completely supplanted for visual beings by even relatively minimal 400nm-700nm light, (b) the black light between visual beings and astronomical objects (that are light sources) is easily completely supplanted for visual beings, as demonstrated by how we can see, with unaided vision and aided vision, and without supplemental light

sources, a profusion of astronomical objects that are light sources, (c) due to *(a)* and *(b)*, it can be inferred that the macrolimit is of black matter and black light, and of intensities that are not easily supplanted by 400nm-700nm light for visual beings, and (d) due to *(c)*, it can be inferred that the removal of a portion of the macrolimit would result in visual beings experiencing that portion of their visual-fields as a region of blindness.

(5) If hypothetically a geometric form comprised the entirety of the observable universe, the removal of a portion of its macrolimit would result in that portion being experienced by observers within the geometric form as a region of sensory-deprivation (that is, a region that is devoid of sensory-stimuli); and as such, the region would be experienced as a region of blindness.

(6) A semblance of the fundamental composition of a geometric form is reached when the result of the dividing, disintegration, and super-resolution microscopy of the conceived of elementary particles within the form is the reappearance of black light.

(7) There is not an other side of the microlimit of a geometric form in the same way that there is of its macrolimit. The other side of the macrolimit of a geometric form is a realm that is outside of the geometric form, whereas the other side of the microlimit of a geometric form is still within the geometric form. All of the space (that is, black matter) of a geometric form, including the space within all of the existents of the geometric form, is the microlimit of the geometric form; and all that can be conceived of, of what exceeds the microlimit, is that it caused the microlimit, and, prior to that, served as the realm for the microlimit to subsist. I will refer to what exceeds the microlimit of a geometric form as the "micro external universe"; and I will likewise refer to what exceeds the macrolimit of a geometric form the "macro external universe."

(8) The possibility of the existence of a geometric form (that is, its microlimit, macrolimit, and content) is the micro external universe.

(9) The micro external universe of a geometric form, since it of course is three-dimensional, and extends to the macrolimit of the geometric form, surely extends into the macro external universe; and as such, and as was speculated earlier, the micro external universe and macro external universe are indeed adjoined; and for this reason, it could be stated that both the micro external universe and macro external universe serve as the possibility of the existence of a geometric form.

(10) Since the microlimit of a geometric form, while *in* the geometric form, is not *of* the geometric form, there may be a tendency to conceive of it as being the fundamental microlimit of the geometric form. However, since black matter and black light are in the geometric form, and since black matter is the microlimit of the geometric form, there must be a realm that exceeds the microlimit by virtue of it being a limit, as was discussed previously; and the realm must be, as was discussed previously, formally different in nature—that is, of a formally different physics.

(11) Regarding *(10)*, in what way, then, is the microlimit of a geometric form a limit? That is, in what way can a three-dimensional attribute that pervades the entirety of a geometric form be a spatial limit? Consider the hypothetical circumstance of super-resolution microscopy reaching pure black light (that is, light that does not consist of any relative existents), and an observer who is physically smaller than all of the relative existents that were passed by, by the microscopy, on its way to pure black light. And let us grant that the observer's visual system is such that the observer can see only the realm of the observer's size—that is, the realm of pure black light. And let us also grant that the observer can travel any distance. The entirety of the observable universe for the observer, regardless of where the observer looked, and where the observer traveled, would be one of pure black light: While

the observer, in traveling any distance, would pass through relative existents of the observable universe, the observer would not experience this: The observer could only infer that the observer is, at any particular time, within a particular relative existent by virtue of experiencing being moved from the observer's location to another location. Aside, the above hypothetical was not one in which microscopy reached black matter, and an observer subsisted in that realm: While this is the microlimit of the geometric form, it, as discussed previously, is formally unobservable, intangible, unmeasurable, and inconceivable; and as such, the hypothetical would have been immensely incoherent. Aside, (a) the observer would not experience if the observer entered the realm of black matter, and (b), and notwithstanding, the observer, due to being of a formally different nature than that of the realm of the microlimit, would, I predict, perish in some way upon entering that realm, such as by being compressed, disintegrated, evaporated, melted, subjected to grave nervous-system devastation, etc.

Therefore, the microlimit of a geometric form is a realm that is located at a particular micro-depth within the geometric form; and the realm is located within all observable locations of the geometric form. The microlimit of a geometric form may therefore be referred to as a "micro-depth limit"; and it, as such, is not a boundary-limit, as is the case of the macrolimit of a geometric form.

(12) Due to how the microlimit of a geometric form is of a formally different nature, I predict that at any level of super-resolution microscopy beyond the initial level of the observation of pure black light, only black light would be observed. While black matter may be reached—though we would not know this—it, due to being formally different, cannot be observed. We are perpetually blind to that realm, and would continue to observe only black light. Black light is the only aspect of the microlimit that is imparted into the geometric form that observers can observe; and while the microlimit surely consist of more than black, we, with our visual-systems, would only see black. Aside, I speculate that if the micro external universe was reached

by super-resolution microscopy—though, again, we would not know this—we would observe what we do observe as a region of blindness.

(13) Regarding where the micro external universe of a geometric form discontinues: As was discussed previously, all aspects of the external universe are formally different than those of the geometric form, except for the geometric form's black matter and black light, which, again, while they are in the geometric form, they are not of the geometric form. As such, questions regarding the origin, macro and micro boundaries, causal system (or causal systems), composition, etc., of the external universe, and of black matter and black light, are themselves (the questions) inapplicable. It is not the case, though, that, of the external universe, there was an origin, there are macro and micro boundaries, there is one or more causal systems, and there is a particular composition, but that we are formally constrained from observing, measuring, and conceiving of what they are. Rather, the phenomena of origin, causal system or causal systems, macro and micro boundaries, and composition, are not phenomena of the external universe. That is, it is not the case that there was an origin, and it is not the case that there was not an origin; and it is not the case that there is a composition, and it is not the case that there is not a composition. All that can be accurately thought is that there is an external universe, and that it consists of formally different phenomena. (Thinking that it exists, and that it consists of phenomena, is accurate, because the states of existence and phenomena apply to it, as does the quality of a state: It would not be accurate to think that the external universe neither exists, nor does not exist, that it neither consists of phenomena, nor does not consist of phenomena, and that it neither consists of states, nor does not consist of states: There are highly general aspects of it that it does have).

(14) Regarding where the micro external universe of a geometric form begins; and again, while the microlimit of a geometric form is *of* the micro external universe, it is not *in* the micro external universe: All that can be accurately thought is that it begins (or ends) on the other

side of the microlimit of the geometric form—that is, that it begins (or ends) at a particular micro-depth.

(15) It may be thought that the macrolimit of the micro external universe of a geometric form is where it meets the microlimit of the geometric form. This issue is discussed in *(16)* below in the context of the issue of the relationship between the micro external universe and the microlimit of a geometric form.

(16) Regarding the relationship between the micro external universe and the microlimit of a geometric form: As was discussed about the micro external universe, all that can be accurately thought is that there is a relationship. Aside, and the following is a revision of what I previously discussed: Regarding what the physics is of the micro external universe, and of black matter, and black light, they, since they are of the external universe, neither have a physics, nor do not have a physics. Having a physics is not a highly general aspect of everything, as the micro external universe, black matter, and black light, are neither physical, nor not physical: Physicality, origin, causation, composition, microlimit and macrolimit, etc., are not applicable to the micro external universe, nor what is of the micro external universe: The micro external universe, black matter, and black light, are formally different than anything of the geometric form.

(17) Regarding the relationship between the micro limit of a geometric form and the geometric form: All that can be accurately thought, as was discussed, is that there is a relationship.

(18) The macrolimit of a geometric form—and the form, hypothetically, comprises the entirety of the observable universe—is a boundary-limit, and may be referred to as the "macro-depth limit." Given that it is black, as was discussed earlier, there could be two kinds of macrolimits of a geometric form: (a) It could be the location of where black matter discontinues, and not of a different structure, nor of a different phenomena; or (b) it could be black matter that is structured differently,

or different in phenomena. Given what was discussed earlier, it is black matter that is structured differently, or different in phenomena.

That the macrolimit of the above geometric form is a boundary-limit, it could be exceeded by the kind or kinds of things that could exceed it, such as black matter and black light, which would entail that they exit the geometric form.

(19) Regarding what the cause is of a geometric form: The first question, and most fundamental question, is what caused the condition for there to be the geometric form.

A prevailing concept of the origin of not only the observable universe, but of the entirety of the universe, is that it (the entirety of the universe) emerged from an infinitesimal conglomeration of matter; (and by "infinitesimal," what is meant is a size of such minuteness that it is immensely beyond even the possibility of observation, measurement, and coherent conceptualization, and that it can only be represented with an austere number that is mathematically-mechanically generated, and which, as such, again, can only be highly vaguely conceived of—that is, conceived of in relation to the intricacy of the observable universe). An aspect of this concept of the origin of the universe is that the universe, itself, expanded—that is, that it did not expand into preexisting space, but rather, that it itself expanded, and that on the other side of its expanding macrolimit, there was nothing (that is, there was no space), including as it expanded. However, while there is a coherent explanation for how particular conglomerations of matter, variations of matter, and densities of matter, can proliferate in size, and over distance, there is no explanation for how the macrolimit of the aforementioned geometric form (that is, the geometric form whose interior consisted of the infinitesimal conglomeration of matter, and which, at that time, comprised the entirety of the universe) could expand: What is conceived of is that the entirety of the universe, at the time of the infinitesimal conglomeration of matter, was the infinitesimal conglomeration of matter; and it would be necessary to demonstrate, via coherent conceptualization, how something, such as an infinitesimal conglomeration of matter itself, can expand into where there is no where. The only even minimally coherent explanation is that the infinitesimal conglomeration of matter—that is, the geometric form—was not the entirety of the universe, nor the entirety of the observable universe, and that, instead, there was a preexisting observable universe into which it expanded. And as for the causal condition of that geometric form itself, there is not an even minimally coherent explanation—that is, there is not even a minimally coherent explanation for how the geometric form (a) arose into existence, (b) arose into existence where, previous to it arising into existence, there was no where, and (c) did not have a prior state. And (d), it would moreover be necessary to explain how a geometric form could exist without having a prior state.

(20) For the hypothetical of a geometric form comprising the entirety of the observable universe: Due to how the macrolimit of a geometric form is of a formally different nature, I predict that at any level of telescopy, only black light would be observed. While black matter may be reached via telescopy—though we would not know this—it, due to being formally different, cannot be observed. We are perpetually blind to that realm, and would continue to observe only black light. As discussed in (18), black light is the only aspect of the macrolimit that is imparted into the geometric form that observers can observe; and while the macrolimit surely consist of more than black, we, with our visual-systems, would only see black. Aside, I speculate that if the macro external universe was reached by telescopy—though, again, we would not know this—we would observe what we do observe as a region of blindness.

(21) Given that the microlimit (that is, black matter; and both, as discussed, are space) of a geometric form (a) is not of the geometric form, (b) was caused by the micro external universe, and as such, underwent a cause that cannot be explained, replicated, nor even conceived of, and (c) cannot be observed, measured, nor conceived of, there is permanent uncertainty about (a) when it arose, (b) how it arose,

(c) why it arose, (d) the size of it when it arose, (e) if it changed in size thereafter, and if so, how it did so, and (f) the relation between it and the other interior of the geometric form.

(22) Only the aforementioned hypotheticals about the macrolimit of the geometric form were adduced, because, as was discussed, no geometric forms that comprise the interior of the observable universe have macrolimits. That is, the geometric forms are only relative forms (or relative existents), and are not discrete existents, because in order for there to be discrete existents, there would have to be macro external universe between them, which is not the case of the observable universe.

91. Regarding what I refer to as black matter, it is clearly the case, from the aforementioned discussions, that black matter, notwithstanding that it is formally different than anything *of* the observable universe, is not accurately characterized as being matter. There surely are one or more alternatives to matter; and we are formally incapable of conceiving of them.

Regarding what I refer to as black light, the above is not entirely the case: Unlike what is the case with black matter, we can observe black light with unaided vision and aided vision; and we can moreover observe relatively more pure states of black light; and we moreover can coherently conceptualize reaching pure black light via super-resolution microscopy; and we moreover can coherently conceptualize pure black light. Moreover, notwithstanding that black light cannot be precisely observed, measured, and conceived of—that is, observed, measured, and conceived of as what it itself is—it can be coherently conceived of as light, since it is visually observed, and both with unaided vision and aided vision. Moreover, it can be coherently conceived of as visible-light, and, moreover, as the only visible-light in the observable universe, as was discussed previously. Moreover, since black light is partly supplanted by 400nm-700nm light (which I have referred to as "common visible-light" and "true non-visible light"), this surely means that it itself is also light: If it was not light, it would not be partly supplanted by another kind of light. Moreover, regarding black light being partly supplanted by "another kind of light," I will predict that since it is partly supplanted by 400nm-700nm light, it itself is the same *kind* of light, but different in variation, similar to how 400nm-700nm light consists of a multitude of variations of light, which are currently characterized as being of particular wavelengths.

Perhaps therefore black light is not formally different than everything that is *of* the observable universe; and perhaps, as such, it, and different intensities of it, could be measured as, but not in the same way as, other variations of light are measured. Aside, and as is well established, there is uncertainty even about what measurable light is itself; and as such, the prospect of being able to, in addition to measuring black light, observing what it is itself seems futile. Also aside, regarding the measurement of light, the measurement technique does not establish what light is itself, nor even that it has the form of a wave.

92. Professor Sean Carroll states the following about what is referred to as "dark matter" and "dark energy"; and as I will discuss shortly, much of his discussion (and he advances a prevailing discussion) is consistent with my discussion; although we of course arrive at our discussions in considerably different ways.

> .... ordinary matter, including all of the elementary particles we've ever detected in laboratory experiments, only makes up about 5% of the energy density of the universe. The rest, of course, comes in the form of a dark sector: some form of energy density that can be reliably inferred through the gravitational fields it creates, but which we haven't been able to make or touch directly ourselves.
>
> .... it's natural to wonder whether the dark sector might be complicated, with a rich phenomenology all its own.

> …. the dark sector …. is dark matter, 25% of the universe, which we know is like "matter" because it behaves that way—in particular, it clumps together under the force of gravity, and its energy density dilutes away as the universe expands. And then there is dark energy, 70% of the universe, which seems to be eerily uniform—smoothly distributed through space, and persistent (non-diluting) through time.
>
> We're imagining there is a completely new kind of photon, which couples to dark matter but not to ordinary matter. …. "dark photons."
>
> What we are proposing is that the dark matter is really a *plasma* …. [11]
>
> 95% of the universe …. is completely invisible. …. it comes in two components: 25% "dark matter" …. and 70% "dark energy" …. [12]

In an article by CERN about dark matter and dark energy, the following is stated about dark energy, which I adduce in order to supplement Carroll's above discussion:

> Dark energy makes up approximately 68% of the universe and appears to be associated with the vacuum in space. It is distributed evenly throughout the universe, not only in space but also in time—in other words, its effect is not diluted as the universe expands. The even distribution means that dark energy does not have any local gravitational effects, but rather a global effect on the universe as a whole. This leads to a repulsive force, which

11. Sean Carroll, "Dark Photons," *Discover Magazine,* www.discovermagazine.com/the-sciences/dark-photons, October 30, 2008.
12. Sean Carroll, "The Preposterous Universe," www.preposterousuniverse.com/preposterous.html, 2008.

> tends to accelerate the expansion of the universe. The rate of expansion and its acceleration can be measured by observations based on the Hubble law. These measurements, together with other scientific data, have confirmed the existence of dark energy and provide an estimate of just how much of this mysterious substance exists.[13]

In the aforementioned article by CERN, the following is stated about dark matter; and as I will discuss below, much of this discussion is consistent with my discussion; although our discussions are arrived at in considerably different ways.

> Galaxies in our universe seem to be achieving an impossible feat. They are rotating with such speed that the gravity generated by their observable matter could not possibly hold them together; they should have torn themselves apart long ago. The same is true of galaxies in clusters, which leads scientists to believe that something we cannot see is at work. They think something we have yet to detect directly is giving these galaxies extra mass, generating the extra gravity they need to stay intact. This strange and unknown matter was called "dark matter" since it is not visible.
>
> Unlike normal matter, dark matter does not interact with the electromagnetic force. This means it does not absorb, reflect or emit light, making it extremely hard to spot. In fact, researchers have been able to infer the existence of dark matter only from the gravitational effect it seems to have on visible matter. Dark matter seems to outweigh visible matter roughly six to one, making up about 27% of the universe.
>
> Many theories say the dark matter particles would be light enough to be produced at the LHC. If they were created at

13. CERN, "Dark Matter," https://home.cern/science/physics/dark-matter, 2020.

> the LHC, they would escape through the detectors unnoticed. However, they would carry away energy and momentum, so physicists could infer their existence from the amount of energy and momentum "missing" after a collision. .... One theory suggests the existence of a "Hidden Valley," a parallel world made of dark matter having very little in common with matter we know.[14]

Via the observation of black light, optical analyses, geometric-physics analyses, mathematical analyses, and related conceptualization, I argue that the microlimit of the observable universe is black matter, and that what is currently thought of as "space" is, actually, (a) relative space, in that it is replete with matter, and (b) the microlimit of the observable universe. I moreover argue that black matter is what causes gravity, in both the micro observable universe, and the macro observable universe. I moreover argue that black matter emits black light, and that both are pervasive throughout the entirety of the observable universe. I moreover argue that different extents of the condensation of black matter are what cause the different extents of gravity in the observable universe.

Carroll, CERN, et al., begin their discussions via the inference of dark matter from their observation that there must be an underlying matter that causes the gravity that holds rotating macro-systems of the observable universe together—that is, gravity without which the rotating systems would have never formed, and currently without which they would unravel from being rotating systems—that is, they would discontinue being rotating systems, and any other kind of systems, via the dispersion of their content. Carroll moreover argues that dark matter is "like matter" "because it behaves that way": "in particular, it clumps together under the force of gravity, and its energy density dilutes away as the universe expands." As such—and notwithstanding, as he concedes, that dark matter has never been observed, measured, nor

14. Ibid.

created—he adduces that it "clumps together," implying that it can be inferred to be clumped together more in regions in which gravity holds rotating macro-systems together, and less in regions in which there are no rotating macro-systems. He moreover states that he speculates that dark matter is of the form of plasma; and as an aside, and as such, when he states that dark matter is "like matter," he really means that he speculates that it *is* matter; and again, he speculates that it is matter because, as he states, "it behaves that way." He, CERN, et al., moreover argue that dark matter comprises about 25% of the observable universe; and they additionally argue that of the remaining approximate 75% of the observable universe, about 70% is comprised of dark energy, and about 5% is comprised of observable matter (which of course includes matter which, while not visually observable, can be measured).

In contrast, I demonstrate that black matter (a) while matter, is a formally different kind of matter than any matter *of* the observable universe, (b) is not of the observable universe, (c) is black, (d) pervades the entirety of the observable universe, including within all observable matter, (e) holds the content of each relative existent of the observable universe together, such that they are relative existents: without black matter, relative existents would disintegrate, (f) is condensed to different extents throughout the observable universe: where gravity is stronger, it is more condensed; and where gravity is weaker, it is less condensed, (g) causes gravity in the sense of the effect that its different extents of condensation have on observable matter, (h) serves as the underlying mechanism that spins, and orbits (that is, propels in orbit, and sustains in orbit), etc., all of the relative micro-existents and relative macro-existents that seemingly have innate spins, orbits, and other mechanical characteristics, and (i) emits black light.

Regarding *(c)*, I added this argument, namely that black matter is black, because I speculate that the reason that black light is black is that its source is also black. I also added *(e)* because, in addition to black matter serving as the aforementioned underlying mechanism of the observable universe, and as, as such, the "gravity" that allows the innate mechanics of the micro observable universe and macro observable universe to proceed without unraveling, it also surely holds the content of existents together.

Aside, I will therefore clarify my discussion about gravity by stating that it is *both* the effect of the condensation of black matter, and the effect of the underlying mechanism of black matter. If gravity was only the effect of the condensation of black matter, then if, hypothetically, the black matter above the surface of the earth was immensely condensed, existents would be permanently suspended in it, akin to being suspended in a highly condensed liquid, or within a solid. My previous concept was that the greater the condensation of space (black matter), the greater the gravity; but the aforementioned shows that the opposite is the case: In regions of space with profoundly condensed space, there could be no gravity: Existents could be permanently suspended in place in space. Therefore, gravity is also the effect of the mechanism of black matter in particular regions. As was discussed, black matter spins, and orbits (that is, propels in orbit, and sustains in orbit), all of the relative micro-existents and relative macro-existents that seemingly have innate spins, orbits, and other mechanical characteristics. As such, existents are either partly or wholly prevented from leaving the surfaces on which they rest because the mechanism of black matter either permanently keeps them on the surfaces via persistent immense pushing, or prevents them from unassistedly leaving the surface of the earth via comparatively less pushing. In the case of existents on earth, an aspect of the mechanism of black matter is such that black matter persistently travels from above the surface of the earth into the earth; and all existents on the surface of the earth are subject to this effect. Aside, the condensation of black matter is also an aspect of the effect: The black matter that persistently moves from above the surface of the earth into the earth is of a particular level of condensation, such that if it was more condensed than it is, existents would be subject to greater pushing, and if it was less condensed than it is, existents would be subject to less pushing. Hypothetically, whereas the speed of the mechanism of black matter may be the same in two locations, the black matter may be so minimally condensed in one location that there is extremely minimal pushing, and so immensely condensed in the other location that there is extraordinary pushing.

Returning to the discussion about CERN, Carroll, et al.: Carroll advances a prevailing argument that dark matter "clumps together" as "inferred through the gravitational fields it creates." However, he concurrently states that "it clumps together under the force of gravity," which implies that it is subject to the "gravity" of the content of the observable universe: He means that its clumping is caused by gravity that is external from it, namely, presumably, the curvature of the space where it is located. First, the concept of the clumping of dark matter is consistent with my concept of condensed space. Second, the reason that Carroll et al. do not argue that variations of dark matter (that is, different densities of clumps, different sizes of clumps, and it in its non-clumped form ("its energy density dilutes away as the universe expands")), are themselves the cause of the variations in intensity of what we observe and measure to be gravity is that they do not additionally conceive of dark matter as being (a) pervasive throughout the observable universe, and (b) the underlying mechanism of the innate mechanics of the micro observable universe and macro observable universe: As discussed above, without conceiving of black matter as being the underlying mechanism of the innate mechanics of the observable universe, all that remains is that different condensations of black matter hold the content of the universe, including the content of existents, together to different extents: The phenomena of adhesion do not fully explain what we observe and measure to be gravity. Third, the reason that Carroll et al. conceive of dark matter as comprising only approximately 25% of the observable universe is that they do not conceive of it as underlying the emission of black light; and they of course do not observe black light, nor, as such, conceive of it. Moreover, they conceive of dark matter as being distributed in different densities of clumps, different sizes of clumps, and in non-clumped form, only within galaxy systems, and around other astronomical systems and astronomical objects. Moreover, they do not conceive of dark matter as existing within astronomical existents.

As for how Carroll, CERN, et al., conceive of dark matter itself, Carroll speculates that it is a plasma, which of course would entail that it is comprised of particles of particular natures, and that, moreover, the particles undergo particular mechanics. CERN, in their article, *No Sign Of Dark Light From The Higgs Boson*, states that dark matter is comprised of particles, and that it, moreover, has a "seemingly extremely feeble interaction with ordinary matter." CERN then, later in the article, modifies their position by stating that dark matter has "very weak or no interaction with the ordinary matter," which is to adduce the possibility that it actually does not have a "seemingly extremely feeble interaction with ordinary matter."

While in the above articles by CERN and Carroll, they do not discuss the following, and instead attend to other phenomena and issues, I will nevertheless supplement their discussions with the following perspective about dark matter serving as an underlying "scaffolding" which not only, as CERN and Carroll discuss above, hold galaxy systems, and other astronomical systems, together, but which observable matter is constructed on.

> Our results are consistent with predictions of gravitationally induced structure formation, in which the initial, smooth distribution of dark matter collapses into filaments then into clusters, forming a gravitational scaffold into which gas can accumulate, and stars can be built.[15]

> Model simulations of the large-scale distribution of galaxies have long suggested that galaxies form on a filamentary network [or "filamentary scaffold," as stated in the below title] of dark matter. Now gravitational lensing has yielded a look at that network.[16]

15. Richard Massey, et al. "Dark Matter Maps Reveal Cosmic Scaffolding," *Nature,* January 2007.
16. Bertram Schwarzschild, "Three-Dimensional Mapping of Dark Matter Reveals the Expected Filamentary Scaffold," *Physics Today,* March 2007.

> "It's reassuring how well our map confirms the standard theories for structure formation," Massey [Professor Richard Massey] said. He calls dark matter the scaffolding surrounding the assembly sites of stars and galaxies over billions of years.[17]

Aside, while Massey's statement clearly demonstrates his position that stars, for example, are "built" on clusters (or clumps) of dark matter (which he refers to as "gravitational scaffold[s]"), the subsequent two positions that are provided, while likely the same in nature as Massey's position, use language that could be interpreted as expressing that the scaffolding is simply what preexisting astronomical objects and systems "form" themselves on, and "assemble" themselves on. However, I believe that what is meant by "structure formation," and that "galaxies form on a filamentary network [scaffold]," is that the astronomical objects and systems originated on dark matter, and as, as Massey states, collections of gasses, and then continued to be constructed on dark matter.

Also aside, regarding what is meant by "filaments," "clusters," and "clumps" of dark matter, in contrast to its initial state ("the initial, smooth distribution of dark matter"), which is also characterized by Carroll as its dilute state; and Carroll also discusses that it "dilutes away as the universe expands": There may be three interpretations of what is meant by "filaments," "clusters," and "clumps" of dark matter: (a) adjacent conglomerations of once separate regions of dark matter, (b) the combining of, or integrating of, conglomerations of once separate regions of dark matter, such that the conglomerations integrate to increase in size and density, or only density, and (c) condensations of, or "collapses" (Massey) of, preexisting smoothly distributed, or dilute, dark matter. Given the discussions about the preexisting smooth distribution of dark matter, and that it can, from whatever subsequent form it has, be diluted, I believe that only *(b)* and *(c)* are the case, or only

17. European Space Agency, "First 3D Map of the Universe's Dark Matter Scaffolding," July 2007.

(c) (and that *(c)* gives the appearance of *(b)*), and that *(a)* is not the case: If *(a)* was the case, dark matter would still be in its dilute state, or smoothly distributed state; and what is argued above is that the observable universe was constructed upon regions of it that were, and still are, in a non-dilute or non-distributed state.

Also aside, the above *(a)* and *(b)* are consistent with my concept of condensed space (or condensed black matter); although I of course argue that the mechanism of black matter is an additional facet of what we observe, measure, and experience to be gravity. Aside, and related to my above discussion about why CERN, Carroll, et al. do not argue that density-variations of dark matter are themselves the cause of the variations in intensity of what we observe, measure, and experience to be gravity, CERN, Carroll et al. could argue that the adhesive nature of dark matter includes a rubber-like nature, such that movements away from the earth's surface, and away from extensions of the earth's surface, entail extending the local structure of dark matter in such a way that pull-tension in the structure occurs. In this case, gravity would be the pull-tension, as well as the retraction of the pull-tension. (Of course there is pull-tension on the surface of the earth as well). However, there would be no explanation for why an object that traveled from another planet with the same gravity-phenomenon as earth would, as it entered earth's gravitational realm, become subject to earth's gravitation realm, and no longer subject to its original gravitational realm.

In summary: "Gravity," then, is a combination of the aforementioned adhesive phenomenon of black matter, and mechanism of black matter. As such, gravity is not a pull-phenomenon, but rather, a push-phenomenon; and it is the same in nature as the innate spin-phenomena, orbit-phenomena, rotation-phenomena, etc., of the micro observable universe and macro observable universe. And I will postulate that the different kinds of magnetism, and degrees of magnetism, are not pull-phenomena, but rather, push-phenomena that arise from the different kinds and degrees of magnetism increasing the speed at which black matter descends; and in addition to the speed being increased, the amount of black matter that descends could also be increased, such as via a diluting-away of black matter in surrounding areas, and a condensation of the black matter that was collected.

Also in summary, it is clearly therefore the case that black matter substantially interacts with observable matter (that is, ordinary matter), and that it, moreover, intensely interacts with observable matter: (a) In order to serve as the aforementioned scaffolding, black matter would have to substantially and intensely interact with observable matter: The building of galaxies on scaffoldings, or within scaffoldings, could not occur if the scaffoldings had "seemingly extremely feeble interaction with ordinary matter," or "very weak or no interaction with the ordinary matter"; (b) the holding together of the content of galaxies by black matter could not occur if it had very weak or no interaction with observable matter; and (c) the "gravity" and other innate mechanics of the observable universe could not occur if black matter had very weak or no interaction with observable matter.

For clarification: Regarding an aforementioned discussion that I added about the nature of the scaffolding of black matter, scaffoldings are, by nature, structures within which, or on which, certain other matter is constructed. It is not something that a finished system of matter enters into. In the aforementioned case of galaxies, it is of course not the case that finished galaxies then enter into scaffoldings: The galaxies would have not developed in the first place without the scaffoldings.

Also for clarification: Scaffoldings, by nature, interact with what is built upon them: They provide an elaborate surrounding structure that the initial matter of objects, or the initial matter of systems of objects, is constructed upon, and without which the construction would not be possible. Moreover: Unlike most architecture, molded objects, etc., which, after they are made, do not need their scaffoldings any longer, astronomical objects, and systems of astronomical objects, enduringly depend on their scaffoldings, as CERN discusses above. As such, it is not the case that black matter has "seemingly extremely feeble interaction with ordinary matter," or "very weak or no interaction with the ordinary matter."

93. Regarding dark energy: As discussed above, Carroll states the following:

> And then there is dark energy, 70% of the universe, which seems to be eerily uniform—smoothly distributed through space, and persistent (non-diluting) through time.

He partly elaborates on this in his aforementioned article, *The Preposterous Universe*, by stating the following:

> The difference between the two [dark matter and dark energy] is in how they behave: dark matter acts like ordinary particles, in that it collects into dense regions (like galaxies or clusters of galaxies), whereas dark energy is smoothly distributed throughout space and slowly-varying in time.
>
> Dark energy is smoothly distributed, but affects the geometry of spacetime itself: it makes distant galaxies appear to accelerate away from us, and it "flattens" the geometry of space, two effects which have been directly observed.[18]

He describes (a) what dark energy is not, namely that it is not comprised of particles, and does not, as such, have the mechanics of particles, (b) that it is distributed evenly throughout space, and does not dilute over time, and (c) the effect that it has on the observable universe, namely that it caused, and continues to cause, its expansion to be of a greater speed than it was in the distant past. However, the above does not provide a description of even the general form of dark energy. Amid his discussion, though, he does provide the below statement in which he specifies the kind of energy that dark energy may be:

> The best candidate for dark energy is the cosmological constant,

18. Sean Carroll, "The Preposterous Universe," www.preposterousuniverse.com/preposterous.html, 2008

> or "vacuum energy": the idea that there is a nonzero amount of energy density inherent in the fabric of spacetime itself.[19]

Whereas the general form of dark matter, according to Carroll, et al., is that it is comprised of particles, and is the structure on which, or within which, galaxies and other astronomical systems are built, and is the medium that adheres the components of galaxies and other astronomical systems together, the description of the general form of dark energy is comparatively extremely minimal in specificity: It is described as energy, which comprises approximately 70% of the observable universe, and which is distributed evenly, and which causes the observable universe to expand at a greater speed than it did in the distant past. Each aspect of this description is extremely minimal in specificity: (a) "Energy" is simply a postulated property of observable matter, and cannot, itself, be observed, nor measured. (b) He states that dark energy is "inherent in the fabric of spacetime itself"; but this also is simply a postulate. (c) That dark energy comprises approximately 70% of the observable universe was arrived at by referring to the percentage of the observable universe that remains after making general measurements of the observable matter of the observable universe, and general measurements of the inferred dark matter of the observable universe: Dark matter began to be inferred and conceived of in the 1600's, and more rigorously in the early 20th century; and dark energy began to be inferred and conceived of in the 1990's. Moreover, the overall region of the observable universe that is conceived of to consist entirely of dark energy may consist of one or more other properties, or one or more other properties in addition to dark energy. (d) Regarding dark energy being distributed evenly throughout space (that is, throughout the approximate 70% of the observable universe where it is located), this postulate may be due to the inability to observe it, or measure it, precisely: For example, perhaps dark energy, in increasing the speed of the expansion of the observable universe, and, as such, increasing the speed at which dark matter is diluted, undergoes a change in the nature

19. Ibid.

of its distribution in the regions in which dark matter is diluted. Carroll does state that dark energy is "slowly-varying in time," by which he is referring to its change in speed as a whole, and in particular, that it began expanding faster approximately 7.5 billion years ago.[20] However, and again, perhaps the nature of the distribution of dark energy, and the speed at which it extends, vary throughout the observable universe, and in particular, in regions of dark matter, due to its dilution of dark matter: In order for regions of dark matter to be diluted by dark energy, they would have to be affected by dark energy; and this would entail an interaction between dark matter and dark energy; and the interaction would surely affect the nature of the distribution of dark energy (it surely would become more dense), and the speed at which it extends (it surely would slow), and perhaps other aspects of it.

Aside: Regarding that dark energy comprises approximately 70% of the observable universe, and how Carroll states that it is "smoothly distributed throughout space," and that it may be "energy density inherent in the fabric of spacetime itself," space ("the fabric of spacetime") is therefore also a facet of this approximate 70% of the observable universe. That is, it and dark energy comprise the approximate 70%. I will continue this discussion below in the context of a discussion about an aspect of Albert Einstein's concept of space.

To begin: In the aforementioned article by NASA, three general perspectives are provided about what dark energy is. The first is the following:

> dark energy .... is a new kind of dynamical energy fluid or field, something that fills all of space but something whose effect on the expansion of the universe is the opposite of that of matter and normal energy. Some theorists have named this "quintessence," after the fifth element of the Greek philosophers. But, if quintessence is the answer, we still don't know what it is like, what it interacts with, or why it exists. So the mystery continues.

20. NASA, "Dark Energy, Dark Matter," https://science.nasa.gov/astrophysics/focus-areas/what-is-dark-energy, January 2020.

The second is the following:

> Another explanation for how space acquires energy comes from the quantum theory of matter. In this theory, "empty space" is actually full of temporary ("virtual") particles that continually form and then disappear. But when physicists tried to calculate how much energy this would give empty space, the answer came out wrong .... So the mystery continues.

The third is the following:

> .... dark energy .... is a property of space. Albert Einstein was the first person to realize that empty space is not nothing. Space has amazing properties, many of which are just beginning to be understood. The first property that Einstein discovered is that it is possible for more space to come into existence. Then one version of Einstein's gravity theory, the version that contains a cosmological constant, makes a second prediction: "empty space" can possess its own energy. Because this energy is a property of space itself, it would not be diluted as space expands. As more space comes into existence, more of this energy-of-space would appear. As a result, this form of energy would cause the universe to expand faster and faster. Unfortunately, no one understands why the cosmological constant should even be there, much less why it would have exactly the right value to cause the observed acceleration of the universe.

The first two perspectives consist of conceptions of dark energy as something that is separate from space itself, but which imbues space; and the third perspective consists of a conception of dark energy as being an aspect of space itself, and is the conception that Carroll advances. Entailed by the third perspective is that space, itself, expands. That is, "it is possible for more space to come into existence." And to further clarify what is meant by energy being an aspect of space itself: Space "can possess its own energy." Now, regarding the third perspective: Notwithstanding that the following is not explicitly stated, or is perhaps obviated, it clearly appears that space is conceived of as something that expands via the generation of itself, and that it generates itself via its own energy. Again, the following is stated; and the following consists of a series of accretional statements.

> Because this energy is a property of space itself, it would not be diluted as space expands. As more space comes into existence, more of this energy-of-space would appear. As a result, this form of energy would cause the universe to expand faster and faster.

Aside, the general concept that should be expressed is that space expands via the generation of itself, and that it generates itself via it and its own energy. However, in the above statements there is a discord, and as such, an incoherency: As the accretion of the statements proceeds, what is expressed is that (a) space, without its energy, perpetuates itself, and (b) ensuingly its energy appears, which causes the universe to accelerate. The general concept is actually that space and its energy are combined, such that neither would exist without the other; and as such, neither of the above two causal stages for the perpetuation of space is the case.

Notwithstanding the aforementioned clarification of the general concept, the concept is immensely negligent: There is not even a minimal explanation for how space (which, again, is the inextricable combination of space and dark energy, and which, as such, could be referred to as "space-energy") perpetuates itself; and there, of course, are no analogies that could be referred to, as there is nothing in the observable universe that perpetuates itself. Moreover, there is nothing else of the observable universe that is speculated to perpetuate itself. Moreover, in the case of the concept of space perpetuating itself, what is conceived of is the perpetuation of the observable universe itself, rather than the perpetuation of some of the content of the observable universe.

The aforementioned general concept is, moreover, immensely dissociative, because it uses the concept of a kind of matter-perpetuation that is clearly observed and understood, and ignores the causal conditions for that matter-perpetuation. For example, cell-division, the expansion of water into non-water areas, the perpetuation of fire, etc., are clearly observed and understood to have intricate causal conditions that allow for their division, expansion, and perpetuation. Einstein, et al., used this concept and concurrently ignored the necessity of a causal condition; and in its place, he, et al., and without an even minimal explanation, stated that space-energy causes itself. Aside, the statement that X (or X-phenomenon) causes itself (namely, X, or X-phenomenon) is a meaningless what I will refer to as a "serial subtractive linguistic alteration, identity linguistic alteration," which is similar in part to the previously discussed subtractive linguistic alterations: The statement that X is caused by Y is clearly understood, and both because this reflects the nature of the phenomena of the observable universe, and because this reflects an aspect of the formal logic of our thought, which of course is based on the nature of the phenomena of observable universe; and stating that X causes itself, or that X is caused by X, is to simply (a) subtract Y, and (b) exchange the identity of X in its place, namely X. Also aside: The above dissociation is the basis for the above linguistic alterations: The linguistic alterations would not be made without the dissociation; and the dissociation is furtively used to substantiate the linguistic alterations.

The above dissociation, and the variety of ensuing linguistic alterations, are undergone when observational progress, experimental progress, and theoretical progress are stifled; and the dissociation and ensuing linguistic alterations are readily accepted because they are highly creative, highly complex, and psychologically puissant, and because mathematics is generated to support them which, notwithstanding that it also consists of immense dissociation and linguistic alterations, is also highly creative, highly complex, and psychologically puissant.

Now, returning to the aforementioned concept of space-energy, I will first reiterate the following about the aforementioned three perspectives: The first two perspectives consist of conceptions of dark energy as something that is separate from space itself, but which imbues space; and the third perspective consists of a conception of dark energy as being an aspect of space itself, and is the conception that Carroll advances. I will argue that the general concepts of space and energy (that is, dark energy) are conflated via dissociation, and in order to obviate problems that are entailed by conceiving of them separately. As is stated above by NASA:

> Because this energy is a property of space itself, it would not be diluted as space expands.

(And again, NASA means to state that dark energy would not be diluted because (a) it is inextricably combined with space, and (b), and as such, expands with space). By conflating space and dark energy into space-energy, the problem of the dilution of dark energy by ever expanding space (or perpetuating space) can be obviated. The problem that is obviated is that (a) there is a limited extent of dark energy in the observable universe, and (b) as space expands or perpetuates, the preexisting extent of dark energy would expand into the new space, and as such, be diluted.

The above conflation of space and dark energy into not two combined and, as such, interrelated existents, but one existent, is dissociative because energy and space are formally different kinds of existents, and to conceive of them as being one existent is to (a) use the concept of a kind of matter-combination that is clearly observed and understood, and (b) ignore the causal conditions for that matter-combination. For example, different variations of liquid can combine into one liquid, and different variations of gas can combine into one gas. Space and energy are formally different in nature such that they cannot combine into one existent; and to ignore this, and to notwithstanding conceive of them as one existent, is to engage in the aforementioned dissociation, and to, moreover, when expressing the concept, engage in what I will refer to as a "meaningless combinatorial linguistic alteration"; and I would argue that such dissociation, and linguistic alteration, occurs also with the concept of, and linguistic expression of, "space-time."

More specifically, about space-energy: First, a general concept of space is that it is a spatial region in which the content of the observable universe resides; and a general concept of energy is that it is a property of matter which, moreover, can give rise to, and which does give rise to, various mechanics. Space is not conceived of as a physical existent, but rather, a region that houses physical existents; and energy is conceived of as a property of physical existents Now, Einstein et al. *generally* conceive of space as (a) a physical existent (a "fabric"), which (b) has energy; and in this sense, their conception (their general conception) is not dissociative. However, they do not provide an coherent argument that attempts to demonstrate that space is a physical existent. As discussed before, Einstein inaccurately, and dissociatively, argues that space is a physical entity due to how it is curved in places. Aside from how he arrived at this (namely, via the demonstration of gravitational lensing), it is dissociative: Space is conceived of as a particular kind of physical entity; and then the concept of physical entity is ignored (dissociated from); and ensuingly, space is meaninglessly *stated* to be a physical form. More specifically: Space is conceived of as a fabric that is, moreover, sporadically curved; this is only inferred from the above experimentation; the fabric and curvature, themselves, have never been observed nor measured; ensuingly, the concept of space as a fabric of sporadic curvature is ignored, and space is then meaninglessly stated to be a fabric of sporadic curvature. As an illustration: The space of a room is conceived of as partly consisting of a ghost; the ghost's existence is inferred from the otherwise inexplicable presence of air-currents in the room; the ghost, itself, has never been observed nor measured; the concept of the ghost is then dissociated from, and ensuingly the room is meaninglessly stated to in part consist of a ghost.

Due to the above about the attempted conception of space as a physical existent, and because in order for there to be energy there must be an existent (it is the existent that has energy, as energy does not exist in isolation from existents), there is nothing to have dark energy. That is, the above attempted conceptions fail.

In contrast to the above perspectives about dark energy, my concept of dark energy begins with the observation of black light: Dark energy is black energy; and black energy is black light that is emitted from black matter; and black matter is inferred from the observation of, and ensuing conception of, black light. Moreover, the inference to black matter is made from the observation of a physical existent, namely black light. In the case of the above perspectives, dark matter is inferred not from a physical existent, but rather, from physical phenomena, and in particular, secondary physical phenomena (that is, phenomena that is not an aspect of dark matter, as black light is of black matter; and again, black light is an existent, rather than a phenomena): The phenomena that dark matter is inferred from is the adhesion of the intricacy of galaxies and other astronomical systems. While dark matter is inferred to adhere the intricacy of galaxies and other astronomical systems, and while dark matter is inferred to serve as the scaffolding within which galaxies etc are built, galaxies etc are not emitted from dark matter, and are not caused by dark matter. Yes, dark matter is a condition of the existence of galaxies; and galaxies are physically connected to dark matter. However, galaxies are not aspects of the intricacy of dark matter itself. For example, galaxies are not conglomerations of dark matter.

Moreover, regarding the above perspectives about dark energy: While there is a basis to infer that something serves to adhere galaxies, and that something serves as the scaffolding of galaxies, and that something serves to accelerate the expansion of the observable universe, there is nevertheless no basis to infer, with certainty, that what they are is matter and energy, including non-ordinary matter and non-ordinary energy: From the above basis, it could be that what Carroll et al. refer to as the "dark sector" consists of one or more radically different features of the observable universe that would not accurately be described as matter and energy, nor non-ordinary matter and non-ordinary energy; and as an aside, recall that CERN, Carroll, et al., imply that dark matter is non-ordinary matter; (and I assume that they conceive of dark energy as non-ordinary energy):

> .... physicists know very little about dark matter, due to its seemingly extremely feeble interaction with ordinary matter.[21]

> One of the distinguishing characteristics of the dark sector is that it has very weak or no interaction with the ordinary matter ....[22]

> .... dark matter acts like ordinary particles in that it collects into dense regions (like galaxies or clusters of galaxies) ....[23]

> We're imagining there is a completely new kind of photon, which couples to dark matter but not to ordinary matter.[24]

I would now like to further discuss the previously discussed concept of the accelerated expansion of the observable universe via dark energy; and in that context, I would moreover like to discuss the concept of the expansion of the observable universe via proliferating space. First, regarding the concept of the expansion of the universe via proliferating space, it is not the case that what is conceived of is the following: (a) that a semblance of the entirety of space is expanded, analogous to a semblance of the entirety of a body of water expanding, because this would entail immense dilution of the entirety of the space; nor (b) that the entire macrolimit region of space of the observable universe—a macrolimit region that is of a comparatively small size—is expanding, while the rest of the space of the observable universe is not expanding. Rather, what is conceived of is, as NASA states that Einstein argued, "it is possible for more space to come into existence":

---

21. CERN, "Dark Matter," https://home.cern/science/physics/dark-matter, 2020.
22. CERN-CMS, "No Sign Of Dark Light From The Higgs Boson," https://cms.cern/news/no-sign-dark-light-higgs-boson, 2016.
23. Carroll, Sean, "Dark Photons," *Discover Magazine,* www.discovermagazine.com/the-sciences/dark-photons, October 30, 2008.
24. Ibid.

> The first property that Einstein discovered is that it is possible for more space to come into existence.

And the following statement by NASA about Einstein's perspective provides an elaboration of the process of proliferating space:

> Then one version of Einstein's gravity theory, the version that contains a cosmological constant, makes a second prediction: "empty space" can possess its own energy. Because this energy is a property of space itself, it would not be diluted as space expands. As more space comes into existence, more of this energy-of-space would appear.

What is meant is that entirely new space arises into existence: space which is, as such, *extended* from preexisting space, and which is not *expanded* from preexisting space. What is moreover meant is that new space is what first arises into existence, and that thereafter, additional energy (the energy of the new space) arises. Of course it is the case, though, that there is a physical connection between the new space and preexisting space: the new space is extended from the preexisting space. However, there is no explicit explanation about how the new space arises. That is, there is no explicit explanation about what causes the new space. For example, does preexisting space, at the relative edge of the macrolimit of the observable universe, replicate itself? Yes, as was discussed previously: As I stated, "Notwithstanding that the following is not explicitly stated, or is perhaps obviated, it clearly appears that space is conceived of as something that expands via the generation of itself, and that it generates itself via its own energy." Inconsistently, though, NASA states the following about Einstein's perspective:

> .... this form of energy ["energy-of-space"] would cause the universe to expand faster and faster.

The statement expresses that the causal phenomenon is that of preexisting space *expanding* via its energy (dark energy), and not being *extended* by new space; and as I discussed above, the expansion phenomenon is considerably different than the extension phenomenon. What should have been stated is "this form of energy would cause the universe to extend faster and faster."

Aside from the above discussion, I discussed earlier that it is immensely incoherent to conceive of the observable universe extending itself where, previous to any moment of extension, there is not only no space, but no reality (no universe)—that is, extending itself where there is no where. Einstein et al. would have to argue not that space replicates itself via its energy, but that reality replicates itself in some way. This, however, would be equally immensely incoherent. The only way that it could possibly be coherent is that if it was, in the first place, observed, after which attempts at explaining it could be attempted. But, and consistent with my earlier discussion, if it was observed, this would not mean that new space, new reality, etc., was created, but rather, either that another kind of phenomenon occurred within the observable universe, or that the macrolimit of the observable universe was caused to move outward by the external universe (the realm of the universe that is outside of the macrolimit of the observable universe), which of course would entail that the external universe caused additional observable universe: The macrolimit of the observable universe would move outward in the sense of the external universe receding, and subsequently causing additional observable universe in the region from where it receded.

It is therefore the case, consistent with what was also discussed earlier, that dark energy does not accelerate what I will now refer to as the "extension of the observable universe." That is, the measurement of the movement of distant galaxies does not demonstrate that dark energy accelerates the extension of the observable universe. Rather, what the measurement demonstrates is that the *content* of the observable universe, not the observable universe itself, is (a) *dispersing*, and (b) dispersing faster than it did in the past. And (c), the accelerated dispersion is caused by dark energy.

94. Since, as discussed earlier, the microlimit of the observable universe is pervasive throughout the entirety of the observable universe, and since black matter is the microlimit of the observable universe, it is therefore not the case that dark matter (which hereafter I will refer to as "black matter"), comprises approximately 25% of the observable universe. But, it of course does not comprise 100% of the observable universe. Rather, it is dispersed throughout the entirety of the observable universe, and integrated with the entirety of the content of the observable universe; and as such, it would be impossible to estimate the percentage of the observable universe that it comprises.

Similarly, notwithstanding that black light is pervasive throughout the entirety of the observable universe, it does not comprise the entirety of the observable universe.

Rather than black matter, black light, and observable matter comprising different discrete sectors of the observable universe, they comprise different micro-depths of the observable universe, and are enmeshed (that is, integrated).

95. Aside from how it is not the case that the content of the observable universe is divided into the aforementioned percentages, namely approximately 70% dark energy, approximately 25% dark matter, and approximately 5% ordinary matter, and aside from how it is not the case that the observable universe is divided into any discrete regions of content, it is not possible to determine whether the content of the observable universe is comprised of any percentages of general kinds of existents without having observed the macro limit of the observable universe. If hypothetically, of minute beings living within a sphere who cannot observe the macrolimit of the sphere, how could they determine even the approximate percentages of the general kinds of existents of their observable universe?

(93), and earlier discussions, demonstrate that for any selected region of the observable universe, the approximate percentages of the general kinds of existents that comprise the selected region are what I

will refer to as "total volume-percentages." As an analogy, the general kinds of existents of a region of air do not comprise discrete sections of that region, but rather, particular total volume-percentages; and this is because they are integrated. For example, of general existents X, Y, and Z, they, because they are integrated, may comprise approximately 70%, 25%, and 5% respectively of the total volume of the air.

96. If, according to Carroll, and NASA's statements about Einstein's perspective, space is a fabric which, moreover, is integrated with dark energy, there should be an estimate about how much of the aforementioned approximate 70% of the observable universe consists of dark energy, and how much consists of space: Since both dark matter and the fabric of space comprise the approximate 70%, each should comprise, as discussed in *(94)*, a total volume-percentage of the 70% that is less than 70%.

Aside, in an article by NASA (the department of the Chandra X-Ray Observatory), *A Tour of the Dark Energy Quasar Survey*, NASA states the following:

> .... dark energy is a proposed type of force, or energy, that permeates all space and causes the expansion of the Universe to accelerate.[25]

This is another expression of the aforementioned concept of Carroll, Einstein, et al., that space and dark energy are integrated into one form, and, that space (a) expands, and (b) has its expansion accelerated by its dark energy.

97. Regarding the concept of space-energy of Carroll, Einstein, et al., recall that the following is stated about the integration of space and dark energy into one form:

25. NASA, "A Tour of the Dark Energy Quasar Survey," Chandra X-Ray Observatory, Harvard-Smithsonian Center for Astrophysics, www.chandra.harvard.edu/resources/podcasts/ts/ts290119_hd.html, February 15, 2019.

> As more space comes into existence, more of this energy-of-space would appear.

This expresses that there is a causal relationship between space and dark energy, namely that, firstly, space expands, secondly, dark energy appears in the expanded space, and thirdly, dark energy accelerates the expanding of the expanded space. As such, this is contrary to the concept of space-energy: It would not be contrary if it was argued that space and dark energy, simultaneously, "expand," and concurrently accelerate. Moreover, that it is argued that dark energy accelerates the expansion of space, and not that dark energy and space accelerate together, demonstrates that what is being conceive of is not space-energy.

98. My discussions to this point demonstrate that the space of the observable universe (a) does not expand, (b) has never expanded, (c) does not extend itself, and (d) has never extended itself. Moreover, the content of the observable universe (a) does not expand, and (b) has never expanded. Instead, the content of the observable universe (a) has dispersed, and (b) at different rates of acceleration; and (c) it is dispersing, and (d) at a different rate of acceleration than it did in the distant past.

99. An explanation that is superior to the aforementioned prevailing explanations, and one that is, moreover, consistent with the geometric-physics that I discussed earlier, is that what is commonly observed and conceived of to be space (namely, the microlimit of the observable universe, which is black matter) is (a) preexisting—that is, it existed prior to presence of any of the "ordinary matter" content of the observable universe, (b) is the realm in which ordinary matter disperses, (c) is not, itself, changing in expansiveness, and (d) is caused by what exceeds it, namely what exceeds its microlimit and what exceeds its macrolimit, namely the two regions of the external universe.

A likewise superior explanation to the explanation of string-theory that gravity is caused by phenomena that occur in "hidden extra dimensions" is the following: The pervasiveness of the microlimit of the observable universe (that is, black matter) throughout the entirety of the observable universe (including within all ordinary matter), at a particular micro-depth that is located within all observable locations of the observable universe, is the cause of micro-gravity and macro-gravity, via the different extents of the condensation of black matter, and via the different mechanistic processes of black matter.

100. In a previously discussed article by CERN, namely *Dark Matter*, the following is stated about the location of dark matter, and about the nature of the interaction between dark matter and ordinary matter:

> One theory suggests the existence of a "Hidden Valley," a parallel world made of dark matter having very little in common with matter we know.

The concept of the location of dark matter as being a "Hidden Valley" is to some extent consistent with my argument that dark matter is located at a particular micro-depth of the entirety of the observable universe. (The micro-depth exists within any observable location in the observable universe; and, again, the microlimit of the observable universe is a micro-depth limit, rather than a boundary limit, as is the case of the macrolimit of the observable universe). However, the Hidden Valley is not a "parallel world" (or parallel observable universe), but rather, a micro-depth within the observable universe. Moreover, a parallel observable universe would not subsist *within* the observable universe, but rather, would be *parallel* to the observable universe, meaning that it would be external from the observable universe, and, as I discussed previously, separated by external universe. While the existents and phenomena of the content of the observable universe can subsist parallel to each other, it is incoherent to state that *within* the observable universe there is a parallel universe (of any nature): The observable universe is one universe. Now, stating that there is a parallel observable universe within the observable universe is an instance of what I will refer to as "combinatorial dissociation," meaning that what is actually conceived of is that two relatively separate realms of the observable universe are superimposed, and then delusively believed to be entirely separate (or formally separate) observable universes, and, moreover, to either minimally coexist, or not coexist. What is dissociated from is that the two realms that are superimposed are realms of the one observable universe, and that the superimposition entails an integration of the realms. As an aside, the following analogy is one in which combinatorial dissociation does not occur; and I adduce it in order to further illustrate the nature of the combinatorial dissociation: An inferred parallel earth of the earth can be conceived of to be superimposed over, or within, the earth; and of course, the inferred parallel earth of the earth is unobservable, and unmeasurable. Now, the earth is conceived of as consisting of a parallel earth within itself that is of a radically different nature, and perhaps a formally different nature; and this is coherent, because it is possible: It is not a necessity that there is only one earth: There is no macrolimit surrounding the earth (at perhaps an elevation above the earth), the other side of which is the external universe. If there was, there would not be a parallel earth to be superimposed over, or within, the earth; and it would therefore be necessary to undergo the aforementioned combinatorial dissociation in order believe that there is a parallel earth superimposed over or within the earth. Aside: In addition to the above hypothetical being coherent because there is not, as of necessity, only one earth, there may be earths that are of a formally different nature than the earth, and which could, as such, exist superimposed over or within the earth.

The general concept of a Hidden Valley (hereafter, "hidden valley") is an apposite analogy for my concept of micro-depth. However, CERN's particular concept of a hidden valley provides an inaccurate analogy: The hidden valley is not a superimposed hidden realm (that is, an unobservable, non-tactile, and unmeasurable realm) that subsists at the same micro-depth of the observable, tactile, and measurable realm. Rather, it is a realm that is located at a particular micro-depth; and the particular micro-depth is located within all observable locations in the observable universe. The valley that CERN conceives of is actually a micro-depth valley, rather than a typical topographic valley.

101. Whereas black matter is pervasive throughout the entirety of the observable universe at a particular micro-depth, black light—while pervasive throughout the entirety of the observable universe—is of course is not so only at a particular micro-depth; and this is why the black that is observed in the visual-field is not black matter. Moreover, and as I also discussed, the black of black matter cannot be supplanted to any extent by 400nm-700nm light, because it cannot be reached by 400nm-700nm light of any intensity, whereas black light, at various micro-depths (from zero-depth, or that is, the surface of the visual-field, inward, until black matter is reached) can be supplanted to at least some degree, for visual beings, by 400nm-700nm light. I did discuss, though, that there is a micro-depth at which there is pure black light, and that pure black light, as is the case of black matter, cannot be reached by 400nm-700nm light.

Now, since black light is emitted by black matter, the entirety of black matter is illuminated with black light; and in this sense, there is a micro-depth of black light that is the same as the micro-depth of black matter; and perhaps this micro-depth is the only location where there is pure black light in the context of 400nm-700nm light, and that all increases in micro-depth consist of different degrees of an enmeshing of black light with 400nm-700nm light. However, of course it is the case, as was discussed, that in the absence of 400nm-700nm light, there is pure black light at all micro-depths.

Given that black light can be enmeshed with 400nm-700nm light at all micro-depths except that of black matter, it can be diluted at those micro-depths; and this would affect how it accelerates what I refer to as the dispersion of the observable universe.

Aside, I have discussed that 400nm-700nm light can supplant non-pure black light to various extents, and that 400nm-700nm light can be enmeshed with non-black light to different extents; and I find that it is more accurate to characterize the phenomena as that the enmeshing of 400nm-700nm light with non-pure black light entails supplanting the full expression of black light to different degrees, and that the extent of what I will refer to as "experiential supplanting" (that is, the extent that visual beings experience the supplanting of black light) is considerably dependent on the nature of the visual systems of the beings with vision: It is conceivable that there are visual beings for whom comparatively extreme-intensity 400nm-700nm light succeeds at only minimally experientially supplanting black light; and vice versa.

102. In *(100),* I stated the following; (and below I will attend to my use of the description "inward" in the statement): "... the black of black matter cannot be supplanted to any extent by 400nm-700nm light, because it cannot be reached by 400nm-700nm light of any intensity, whereas black light, at various micro-depths (from zero-depth, or that is, the surface of the visual-field, inward, until black matter is reached) can be supplanted to at least some degree, for visual beings, by 400nm-700nm light." I stated "inward," and not "below," nor "downward," because micro-depth is not a depth of height, but rather, inwardness.

Spatial inwardness is the fourth spatial dimension of the observable universe; and it is the only other spatial dimension of the observable universe. As was discussed earlier, every existent of the observable universe, and every region of the observable universe, is three-dimensional; and within the three-dimensionality is the fourth dimension. The four dimensions are the following: length, width, height, and inward. Moreover, the fourth dimension could be graphically represented by the below arrow-lines, which represent the beginning of inward at the surface of the cube, and the ensuing disappearance of the presence of the line into inward. (The arrow of the arrow-line is not a part of the line, and instead is used in order to show the entry of the line into the inward dimension of the cube).

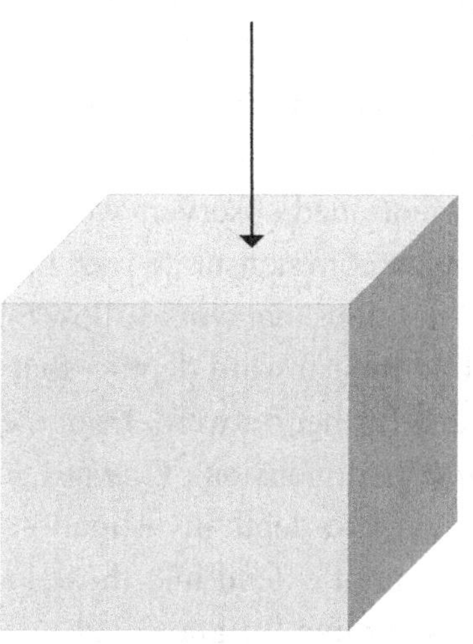

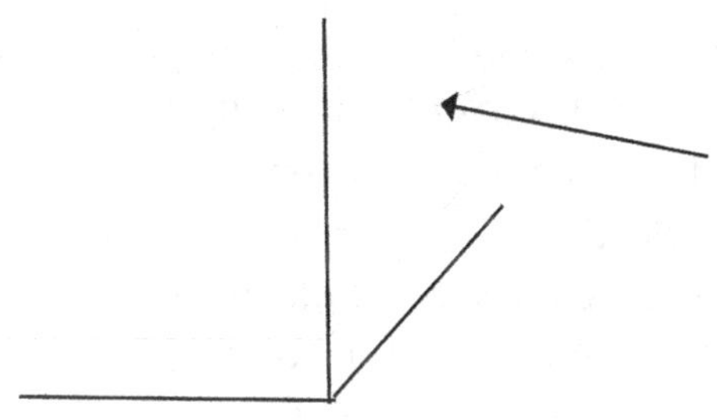

Inwardness is of course not "decrease of size" in the sense of the accretional division of the cube, or a region of the cube. Rather, it is inward depth; and each inward depth extends throughout the entirety of the cube, such that, hypothetically, for any observers at those inward depths, the entirety of cube (of the nature that it is at those micro-depths) can be sensorily and kinetically experienced. Moreover, and to explain what I mean by, "the entirety of the cube (of the nature that it is at those inward depths)": If hypothetically a particular inward depth consisted of particular relative existents, and particular mechanics, phenomena, and colors, that cannot be found anywhere else, those existents, mechanics, phenomena, and colors would pervade the entirety of cube for the observer, and be all that pervades the entirety of the cube for the observer.

Aside, the aforementioned observers would of course have neurologies and bodies that are commensurate with the nature of their inward depths, and lack technologies that would allow them to examine inward depths that are beyond their inward depth—that is, inward depths that are further inward, and further outward, from their depth.

For clarification: The dimension of inward does not extend *into* the above cube; and if the micro-depth investigation begins at any location within the cube, it does not extend into those locations either. Rather, the dimension *disappears* into the above cube.

The aforementioned discussions that I provided about super-resolution microscopy, and micro-depths, are therefore inaccurate, as microscopy is only relevant for the investigation of the micro-intricacy of the zero inward depth (or relative surface) of any region of the observable universe: Microscopy only penetrates the common depth of any relative surface of the observable universe, and does not penetrate inward depth.

I will therefore conclude that inward depth is not an investigatable dimension. But notwithstanding, it is a conceivable one.

Regarding my aforementioned inaccurate discussions about micro-depths, what I conceived of is the investigation of relatively infinitesimal depth-distances of the observable universe, such as the distance between the surface (or relative zero-depth) of any region of the observable universe, and less than the attometer length of 10^{-18}, which is the approximate size of the smallest features of the observable universe that have been measured. I conceived of black matter, and pure black light, as subsisting in a micro-depth realm that is considerably less than 10^{-18}. There are two reasons why these discussions are inaccurate: (1) Since micro-depth realms are of the dimensions of length, width, and height, the entry side of the relatively infinitesimal micro-depth of any region of the observable universe would consist of a macro-boundary. That is, the boundary of the side that is investigated would consist of this macro-boundary. As was demonstrated earlier, the microlimit of the observable universe (again: space, which is black matter) is pervasive throughout the entirety observable universe; and as such, it does not have a macrolimit between it and and the observable universe. (2) If there was a macrolimit between the microlimit of the observable universe and the observable universe, the microlimit of the observable universe would not be pervasive throughout the entirety of the observable universe: There would only be a region of microlimit. And regarding this, I would like to discuss the following about the geometry of the investigation of micro-depth realms. If the microlimit of the observable universe was reached only after, for example a $>10^{-18}$ depth-investigation, the distance between 0 and $>10^{-18}$ would lack a microlimit; and as I discussed earlier, the microlimit pervades the entirety of the observable universe.

For clarification: The inward dimension cannot be coherently conceived to exist alone; and the reasons for this were discussed earlier in the context of the dissociative conceptual separation of the three spatial dimensions of length, width, and height.

103. Regarding micro-depth: Microscopy is not the examination of zero-depth surfaces of existents in the observable universe. (1) As I discussed earlier, there is no such thing as a two-dimensional existent; and as such, and as I stated above, there are only *relatively* zero-depth surfaces. (2) Microscopy is therefore the examination of the micro-depth of existents; and the micro-depth consists of an accretional construction of sub-existents through the depth: Microscopy, like macroscopy (telescopy, etc), enters into those depths.

104. From many of the above discussions, it is evident that there is the question of how ordinary matter and the microlimit of the observable universe are integrated. That is, many of the above discussions imply that ordinary matter is enmeshed with the microlimit: The microlimit, while pervasive throughout the entirety of the observable universe, including within all existents of the observable universe, is formally different in nature than the existents; and as such, ordinary matter and the microlimit are enmeshed. However, this is false, as it would entail that a substantial extent of the observable universe is not comprised of the microlimit: What is conceived of is that ordinary matter is superimposed within the microlimit, and that the superimposed portion of the observable universe is formally different in nature than the the microlimit. Aside, what may also be conceived is that ordinary matter is superimposed in the observable universe, and, is completely separate from the microlimit. However, this conception arises from the aforementioned inaccurate discussions about micro-depth: What is conceived of is that the microlimit subsists at a micro-depth that is three-dimensionally below ordinary matter, however infinitesimal the depth may be.

105. Could ordinary matter be complex conglomerations of the microlimit of the observable universe—that is, complex conglomerations of black matter and black light? Since black matter and black light are not *of* the observable universe, and are formally different in nature than ordinary matter, this could not be the case.

106. The fourth spatial dimension was derived in the way that was discussed above; and its post-derivation legitimacy is found by how it provides a superior alternative to the prevailing perspectives, such as that of the previously discussed extra spatial dimensions of string theory. Aside, and as was discussed, and unlike my discussion about the dimension of inward, Susskind, Randall, et al., do not provide graphical representations of any of the extra spatial dimensions; and they moreover do not provide an even minimal description of them. Instead, they, via meaningless additive linguistic alterations, only state that the dimensions are present, and that they are hidden via being rolled up, curled up, etc. Moreover, Susskind, Randall, et al., provide a supportive analogy with the dissociative conceptualization, and inaccurate neurology, that is conveyed in the novel, *Flatland: A Romance of Many Dimensions*. The analogy is as such highly inaccurate. Moreover, Susskind, Randall, et al., as shown in the previous discussions of theirs that I provided, discuss the extra spatial dimensions as existing independently from the three spatial dimensions of length, width, and height. They describe the dimensions as being hidden, and curled up, and apart from the three spatial dimensions of length, width, and height. They dissociatively argue that the strings of string theory are (a) one-dimensional, and, moreover, (b) exist in the context of extra spatial dimensions, hidden from the context of the three dimensional. They argue that the reason that we cannot observe them via technology (that is, via particle collision experimentation) is that they are located within, and hidden within, the extra dimensions.

107. Black matter and pure black light subsist in the four-dimensional (that is, the dimensions of length, width, height, and inward), and are the cause of the observable universe—that is, they are the cause of (a) ordinary matter, (b) the mechanisms of ordinary matter, (c) the gravity of ordinary matter, (d) the adhering of the content of existents, (e) the adhering of galaxies, (f) the accelerated dispersion of galaxies, (g) the visually critical microlimit of the observable universe, etc.

Pure black light is expressed from the four-dimensional into the three-dimensional, thereby resulting in the presence of non-pure black light in the observable universe. As to why this may be so, I discuss this later.

There of course is no realm that is only of the fourth dimension.

400nm-700nm light of course does not illuminate the four-dimensional. However, the four-dimensional is not "beyond" such light at a particular micro-depth, as I mistakenly thought previously. Rather, it is *apart* from such light in a different dimensional system. If hypothetically 400nm-700nm light entered the dimensional system at a particular region of the observable universe, the light would disappear from the spatial-field.

There is no "fabric" of space, fabric of "space-time," etc., to penetrate. Moreover, since such fabrics would be three-dimensional, because space (the microlimit of the observable universe) is three-dimensional, it would not be possible to penetrate them. That is, in conceiving of three-dimensional space as a three-dimensional fabric, it is conceived of as three-dimensional matter (and the use of "three-dimensional" to qualify space, fabric, and matter is redundant, as each *is* three-dimensional, and cannot be otherwise, and cannot be coherently conceived of to be otherwise); and (a) the penetration of the surface of a three-dimensional fabric simply results in the exploration of the identical internal content of the fabric, and (b) it would not be possible, in any case, to penetrate the three-dimensional fabric of the observable universe, because it would not have any surfaces to penetrate: We are already within the three-dimensional fabric, and cannot, as such, penetrate anything; moreover, it would only be from the location of the external universe that such a fabric could be penetrated—that is, the macrolimit, or macro-boundary, of the fabric would be the macrolimit of the observable universe, and it could as such only be penetrated from the location of the external universe: The only limit or boundary of the three-dimensional fabric would be the macrolimit of the observable universe; and the penetration of it could as such only come from what is external to it. But, in any case, the penetration of it would simply result in the exploration of the identical interior of the fabric. As an aside: Similar to the dissociation that is involved in conceiving of space as being sporadically curved, when space is conceived of as a fabric, what is typically conceived of is a relatively thin slice of three-dimensional space; and the entirety of space is conceived of as being that relatively thin slice: The thin slice is represented as being space, and as being the surface that can be penetrated, and curved. This of course is inaccurate; and the inaccuracy is of dissociation, as it dissociates from (a) how the thin slice of the observable universe is only a thin slice of the observable universe, and not as such the entirety of space, and (b) that there is no such thing, in any case, as a thin slice of the observable universe. As I demonstrated above, and earlier, space (the microlimit of the observable universe) cannot be curved (it is not the kind of existent that can bc curvcd), nor penetrated (we are already within it).

108. Inward entry at any location of the observable universe exceeds the microlimit of the observable universe and enters the micro external universe.

(The microlimit of the observable universe, while in the observable universe, is not of the observable universe).

I previously derived that black matter and pure black light comprise the microlimit of the observable universe, and that it is formally impossible to deduce with even minimal certainty what the micro external universe even in part, and generally, consists of. However, after having derived the fourth spatial dimension, I will deduce that the micro external universe (a) consists of what, for us, is black matter and pure black light, and (b) is of a dimensional system (namely the four dimensions) which, while the entry into the dimensional system via the fourth dimension is clearly conceivable, is itself formally inconceivable—that is, the relation between the four dimensions is formally inconceivable. Moreover, the precise relation between the four-dimensional system and our three-dimensional system is formally inconceivable.

Again, black matter and pure black light subsist in the four-dimensional, and are the cause of the observable universe—that is, they are the cause of (a) ordinary matter, (b) the mechanisms of ordinary matter, (c) the gravity of ordinary matter, (d) the adhering of the content of existents, (e) the adhering of galaxies, (f) the accelerated dispersion of galaxies, (g) the visually critical microlimit of the observable universe, etc.

As for what, then, the microlimit of the observable universe is, I will deduce that it is the beginning of the fourth spatial dimension, and that it, as was discussed, is pervasive throughout the entirety of the observable universe, including in all relative existents of the observable universe. What we observe and experience as relative space is, then, the beginning of the fourth spatial dimension.

What would be the case if hypothetically there was no fourth spatial dimension? That is, if there was no four-dimensional micro external universe. There of course would not be an observable universe: The four-dimensional system created and sustains the observable universe, and is its mechanism. This is in contrast to my previous argument that if hypothetically the observable universe did not have a microlimit, there would be blindness, despite having the ability to see: The visual field, including the surfaces of all existents, would be colorless, and unilluminated, including via black light; and as such, there would be blindness despite having the ability to see. Without the fourth spatial dimension, there would not be blindness, but rather, no three-dimensional surface.

However, surely black matter and pure black light extend from the micro external universe to the microlimit of the observable universe; and as such, my discussions about the microlimit remain. What is added to my discussions about the microlimit is that it—that is, any location of it—is the beginning of the fourth spatial dimension. That is, it is a part of the fourth spatial dimension. That is, it itself, in part, is of the nature of the fourth spatial dimension. What are, for us, black matter and pure black light, are four dimensional; and they, in the micro external universe, extend to the observable universe, and form the microlimit of the observable universe.

What then, if anything, could traverse the microlimit of the observable universe?

> .... from the quantum theory of matter .... "empty space" is actually full of temporary ("virtual") particles that continually form and then disappear.[26]

---

26. NASA, "Dark Energy, Dark Matter," https://science.nasa.gov/astrophysics/focus-areas/what-is-dark-energy, January 2020.

These relative particles do not form and disappear, but rather, are created by the micro external universe, and move from the micro external universe to the observable universe, and then retreat to the micro external universe. There is no disappearance of relative particles, but rather, movement of the relative particles from the three-dimensional system to the four-dimensional system. The "disappearance" is the movement of the relative existents into the four-dimensional system. Moreover, there is no appearance (or inexplicable emerging into existence) of relative existents in the observable universe—that is, this does not occur *in the observable universe*—but rather, they move from the four-dimensional system to the three-dimensional system. The "appearance" is the movement of the relative existents from the four-dimensional system to the three-dimensional system.

A prevailing alternative perspective includes the following concept; and the concept, while not directly related to the above discussion, is analogously related:

> Quantum gravity models predict that space-time is a seething foam of tiny regions where minuscule new dimensions unfurl and then furl back in on themselves, spontaneously appearing and disappearing with inconceivable quickness.[27]

What is analogously conceived of is that quantum particles spontaneously appear and disappear, and via the mechanism of "minuscule new dimensions" that "unfurl and then furl back in on themselves": When the dimensions unfurl, the particles appear; and when the dimensions furl, the particles disappear. However, (a) as was discussed, no even general descriptions of the new dimensions are provided; and no schematic representation of the dimensions are provided, except for, in general, a particle-like geometric form that has a multitude of contours throughout the surrounding surface, (b) the aforementioned

27. NASA, "Quantum Foam," https://science.nasa.gov/science-news/science-at-nasa/2015/31dec_quantumfoam, December 2015.

surface-contours do not demonstrate anything more than a three-dimensional relative particle whose surfaces are contoured, (c) the assertion of "new dimensions" is a meaningless additive linguistic alteration, (d) that the relative particle of *(a)* undergoes furling and unfurling only means that its three-dimensional form changes, akin to a piece of dough being subject to radically different shapes via one's hands, (e) if what it means to "unfurl and then furl back in on themselves" is that the relative particle changes from being compact to less compact, and then back to being compact, this only means that its three-dimensional form has changed, as described in *(d)*, (f) if what it means to "unfurl and then furl back in on themselves" is that the relative particle opens, and then closes, this is incoherent, as relative particles do not open and close: there is nothing to open and close; and this is another instance of dissociation, as what is conceived of is a relative existent that has an opening through which the interior of the existent can be accessed, and that the opening opens and closes: This is an instance of dissociation because relative particles do not have such openings, yet they are conceived of as such: That they do not have such openings is dissociated from. (g) Aside, and ignoring for the moment that there are no such openings, there moreover are no hidden dimensions within such openings—that is, dimensions that perhaps appear through the openings, and then retreat into the openings: Aside from what was discussed above about the "new dimensions," dimensions do not appear and disappear through openings: only some of the content of relative particles may appear and disappear via the openings. Moreover, there is nothing that is formally different inside of the openings; and to conceive of the openings as harboring new dimensions is to merely escalate the extent that there is a deficit of even minimal description, the extent of conceptual vacuity, and the extent of dissociation: (1) the dimensions are not described, (2) the dimensions are not schematically represented, (3) the dimensions are identified with additive linguistic alterations, (4) the dimensions are conceived of as being hidden, (5) the dimensions are conceived of as being such that they can furl (or curl), and unfurl (or uncurl): Again, dimensions are not the kind of thing that can curl and uncurl. Only relative existents can do so. And (6), the relative particles are conceived of as having openings in which the dimensions reside.

The above establishes that there are no relative particles of the "extra dimensions." My descriptions of the fourth spatial dimension, the four-dimensional system, the microlimit of the observable universe, the micro external universe, and black matter and pure black light, and the derivations of each from the evolution of my geometric physics and optics, provides, it seems, a superior explanation for quantum phenomena, such as relative particle appearance and disappearance, gravity, entanglement, etc. (Regarding entanglement: Relative particles that have interacted remain physically connected regardless of the ensuing distance between them; and this is likely explained by how the particles are not discrete, but rather, as discussed, relative particles: The particles, as is the case of all of the relative existents of the observable universe, are actually portions of the observable universe that are completely enmeshed; and particular kinds of interactions between portions of the observable universe results in the portions being inexorably physically connected over any distance. Aside, I argued that the reason that there are no discrete existents of the observable universe is that there would be external universe between them if there were, and that this is not the case. The fact of entanglement provides further proof that the entirety of the observable universe is one inseparable unit: Of course it is the case that after the relative existents interact in a particular way, and then depart from one another, there are, so to speak, one or more trail-connections between them, such that when one relative existent is affected in a particular way, the other relative existent will be affected in a particular way. Aside, I would speculate that the destruction of a partner relative existent would result in the other relative existent encountering one or more changes).

109. Notwithstanding, as was discussed earlier, that the precise relation between the four-dimensional system and our three-dimensional system is formally inconceivable, the following is the case of the general relation between the four-dimensional system and three-dimensional system: Given the nature of the fourth spatial dimension, the content of the observable universe is a three-dimensional surface: Every location of the content of the observable universe is a surface of the four dimensional system that lies inward.

The content of the observable universe as a three-dimensional surface of the inward fourth spatial dimension is its fifth spatial dimension. The fifth spatial dimension is the three-dimensional surface of the inward four-dimensional system.

The fifth spatial dimension can be graphically represented by the gray color within the below cube, and at the relative macrolimits of the cube: The gray is the surface of the four-dimensional system.

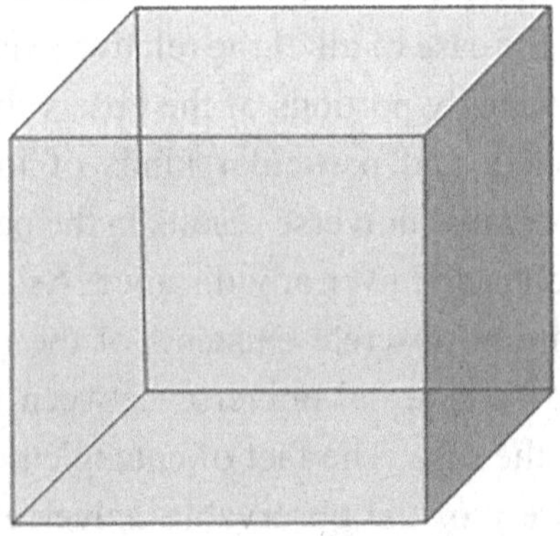

The observable universe, therefore, consists of the first, second, third, and fifth dimensions. The micro external universe consists of the first, second, third, and fourth dimensions. And of course, the dimensions of each dimensional system are inextricably combined. That is, the dimensions of the observable universe are inextricably combined, and the dimensions of the micro external universe are inextricably combined. Aside, I will hereafter refer to the dimensional system of the observable universe as the "1-3|5" dimensional system, and the dimensional system of the micro external universe as the "1-4" dimensional system.

110. Given that the macrolimit of the observable universe would be a boundary limit, unlike the microlimit of the observable universe, and given that there is, as was discussed, a macrolimit of the observable universe, and given that what exceeds it is the macro external universe, and given that it, itself, while it is in the observable universe, is not of the observable universe, and given that it, as such, is an aspect of the macro external universe, and given that, as was mentioned previously, the micro external universe and macro external universe are surely conjoined, I will deduce that traversing the macrolimit of the observable universe entails entering a 1-4 four-dimensional realm, and that the realm is conjoined with the 1-4 four-dimensional micro external universe.

An objection to the deduction of the 1-4 dimensional micro external universe is that it, since it subsists within the 1-3|5 dimensional observable universe, has a 1-3|5 dimensional macro-boundary. However, the deduction of the conjoining of the 1-4 dimensional macro external universe with the 1-4 dimensional micro external universe demonstrates that the micro external universe is not macrolimited by the macrolimit of the observable universe.

Of course, though, the aforementioned two 1-4 dimensional realms, while formally the same, are different in variation, as the micro external universe is within the observable universe, and serves as the cause of (a) ordinary matter, (b) the mechanisms of ordinary matter, (c) the gravity of ordinary matter, (d) the adhering of the content of existents, (e) the adhering of galaxies, (f) the accelerated dispersion of galaxies, (g) the visually critical microlimit of the observable universe, etc., while it is less certain what the nature of the macro external universe is, and, moreover, what its function is for the observable universe, besides providing a visually critical black macrolimit of the observable universe, and a geometrically critical macrolimit, as without a geometric form, there would be no observable universe. Aside, and relatedly, and as is implied by previous discussions, the only existent of the observable universe that has a macrolimit is the observable universe.

I stated that the macro external universe is 1-4 dimensional; and while its four-dimensionality was discussed earlier, I will add that it is at least additionally 1-3 dimensional because it 1-3 dimensionally encompasses the 1-3 dimensionality of the observable universe. That is, the 1-3 dimensionality of the observable universe requires that it is at least 1-3 dimensional.

The macro external universe may of course also consist of the fifth spatial dimension, and one or more additional spatial dimensions. And this of course is also the case of the micro external universe.

111. While I deduced that the macro external universe is, as is the case of the micro external universe, 1-4 dimensional, it surely is also at least 1-3|5 dimensional, as is the case of the observable universe: Outward of the 1-4 dimensionality is surely the presence of a dimensional system (that is, an outward dimensional system), which necessarily is at least minimally 1-3|5 dimensional, as is demonstrated in *(108)* and *(109)*; and while the content of it is of course formally different than that of the observable universe, it surely is the realm that consists of intentional entities that design the inward causal and mechanistic 1-4 dimensional realm.

Perhaps the 1-4 dimensional realm could be conceived of as being slightly analogous to the underlying intricate mechanism of a machine: The machine of the 1-4 dimensional system of the micro external universe is the observable universe; and there surely is a machine of the 1-4 dimensional system of the macro external universe. However, since the machine of the 1-4 dimensional system of the macro external universe is surely what designs the entire 1-4 dimensional system, there surely is another facet of the 1-4 dimensional system of the macro external universe that is formally different than the 1-4 dimensional system of the micro external universe.

112. As is entailed by *(109),* the microlimit of the observable universe is located at the fifth spatial dimension of the observable universe. That is, the microlimit is the three-dimensional surface within the observable universe. That is, the microlimit is the three-dimensional surface of the 1-4 dimensional micro external universe.

Given, as was discussed, that the microlimit of the observable universe is not located at a micro-depth, and, that it pervades the entirety of the observable universe, every location of the observable universe is a three-dimensional surface feature of the micro external universe. That is, the content of the observable universe is a three-dimensional surface feature of the micro external universe. That is, the observable universe is a three-dimensional surface.

What then is the relation between the three-dimensional surface of the observable universe and the micro external universe? The 1-4 dimensionality of the micro external universe is surely causally integrated with the 1-3|5 dimensionality of the observable universe in a particular way that is formally impossible for us to observe, measure, and conceive of. (There is a tendency to conceive of the relation as involving a micro-distance between the 1-4 dimensional micro-limit of the observable universe and the fifth-dimension of the observable universe (that is, the three-dimensional surface of the observable universe), and that the two realms, while perhaps causally related, are separate. However, as was discussed, this is inaccurate). The most that can be conceived of is that the two realms, while formally different in nature, are not separate. And this is consistent with how, as was discussed, it is formally impossible to observe, measure, and conceive of not only the precise nature of the micro external universe, but the causal relation between it and the observable universe. All that can be conceived of is the previously discussed and recently discussed general geometric physics, and general causal physics, all of which is based on, in various ways, or which is consistent with, in various ways, the observation of, the optics of, and the conception of, black light.

113. Regarding what I observe and conceive of to be black light, Carroll, as mentioned, states the following:

> We're imagining there is a completely new kind of photon, which couples to dark matter but not to ordinary matter.

.... particles known (from now on) as "dark photons."[28]

(a) Black matter emits black light. (b) There are only relative photons (relative existents). (c) What is referred to as dark energy is what is referred to as dark photons. (d) What is referred to as dark photons is black light. (e) Carroll et al believe that what they refer to as dark photons are dark in the sense that they, themselves, are formally completely unobservable and unmeasurable. While black light, itself, is formally unobservable and unmeasurable, it can generally be observed (namely, as black light), and its intensities (that is, the extents that it saturates the visual field) can generally be measured. (f) Regarding Carroll's speculation that the "new kind of photon ... couples to dark matter but not to ordinary matter": Again, black matter emits black light (and for the purpose of this discussion, I will refer to black light as "black photons"); due to this, black photons do not couple to black matter; black photons do indeed couple to ordinary matter, namely by illuminating them as black. (As an aside, black matter is black because it emits black light. That is, it is not black because it is illuminated as black by the emission of black light from elsewhere. Also aside, and tangentially: Regarding relative existents in the observable universe that appear as black to different degrees in the context of 400nm-700nm light that supplants the black light illumination of other existents, they would be illuminated beyond black—that is, they would be illuminated as other colors—if the 400nm-700nm light was more intense. Moreover, while they seem to be illuminated by black light to a greater extent than other existents, this is a phenomenon of the human visual system, and, the common 400nm-700nm light-intensities used by humans. What would be relevant is if a portion, or all, of the surfaces of certain relative existents remained black in the presence of comparatively immense extents of 400nm-700nm light: This would demonstrate that black light is emitted to a greater extent by a portion of, or all of, the black matter of the 1-4 dimensional micro external universe of those relative existents).

28. Carroll, Sean, "Dark Photons," *Discover Magazine,* www.discovermagazine.com/the-sciences/dark-photons, October 30, 2008.

114. In the aforementioned article by CERN-CMS (namely, *No Sign of Dark Light from the Higgs Boson*), the following is stated about "dark sector particles," by which of course is meant both dark matter and dark energy:

> In particle detectors like the Compact Muon Solenoid [CMS], dark sector particles produced in the proton collisions are not directly observed. One of the distinguishing characteristics of the dark sector is that it has very weak or no interaction with the ordinary matter of the detector, meaning the particles just pass through the experiment without leaving a signal. Instead, their production can be inferred from the "missing transverse momentum," the imbalance in the sum of particle momenta in the plane perpendicular to the proton beams, which should add up to zero if all particles are detected.[29]

However, how can it be known that there is an imbalance in the sum of particle momenta when it is not known what the totality of the results of the proton collisions are? That is, what would the "missing transverse momentum" be of?: If it is not known what there is at the beginning of the experimentation, it cannot be known what is missing—that is, what passed though the experiment without detection. The question of the issue can also be expressed as the following: If dark sector particles have never been detected, how can it be known that they "pass through the experiment without leaving a signal"? The issue of the question can also be expressed by the following: If "if all particles are detected" "the sum of particle momenta in the plane perpendicular to the proton beams" "should add up to zero," this assumes that all of the particles of the experimentation are known prior to the experimentation; and since dark sector particles are not known prior to the experimentation, the experimentation would be irrelevant for their detection.

29. CERN-CMS, "No Sign Of Dark Light From The Higgs Boson," http://cms.cern/news/no-sign-dark-light-higgs-boson, 2016.

The aforementioned experimentation is more thoroughly explained in an article by CERN, namely, *NA64 Casts Light On Dark Photons.*

> In the new study, the NA64 team looked for dark photons using the missing-energy technique: although dark photons would escape through the NA64 detector unnoticed, they would carry away energy that can be identified by analysing the energy budget of the collisions.[30]

Dark photons, though they themselves are not yet detectable, would, as they pass though the detector, carry away some of the energy of other particles. The missing energy, or missing transverse momentum, has not yet been detected though:

> They [the NA64 team] found no evidence of dark photons in the data.[31]

> The team analysed data collected in 2016, 2017 and 2018, which together corresponded to a whopping hundred billion electrons hitting the target. .... (NA64 is a fixed-target experiment. A beam of particles is fired onto a fixed target to look for particles and phenomena produced by collisions between the beam particles and atomic nuclei in the target).[32]

> The CMS [Compact Muon Solenoid] collaboration … found no signal of dark photons.[33]

---

30. Ana Lopes, "NA64 Casts Light On Dark Photons," CERN, https://home.cern/news/news/physics/na64-casts-light-dark-photons, July 22, 2019.
31. Ibid.
32. Ibid.
33. Ana Lopes, "CMS Hunts For Dark Photons Coming From The Higgs Boson," CERN, www.home.cern/news/news/physics/cms-hunts-dark-photons-coming-higgs-boson, May 24, 2019.

While the expression of black light is observable to human vision, and surely the visual systems of other animals, black light itself has not yet been detected. Since black light in the observable universe is not the pure black light of the micro external universe, I speculate that a measurement method may be devised that detects its effects on other existents, or that a current measurement method may be improved that does this. However, I more strongly speculate that since black light is not of the observable universe, it (that is, its expression) can only be visually observed, and that it itself cannot be non-visually detected. I moreover speculate that its relation to ordinary matter, due again to it not being of the observable universe, will never be observed, measured, nor accurately conceived of.

Since black light accelerates the dispersion of the content of the observable universe, it does interact with ordinary matter.

There clearly is an explanation for why the micro external universe itself is not observable, measurable, and precisely conceivable, and why, instead, all that can be observed is black light, and—when black light is supplanted to various extents for human vision—the 400nm-700nm illumination of the surfaces of relative existents that can be observed by unaided vision and aided vision: (a) Aside from how we would be formally different than we are if we could observe, measure, and accurately conceive of the micro external universe itself, if hypothetically we could do so, our ability to observe and interact with what formerly was the observable universe would be greatly undermined, and surely highly negatively undermined. (b) Black light is the alternative to the daily time-period of sunlight; and it is moreover an alternative that we can choose via shelters of various kinds, and our ability to make, and use as we desire, other light sources. Living amid black light of various intensities, at different time-periods of each day, is neurologically and psychologically crucial, and for an array of obvious reasons. If hypothetically there was no black light of any degree, and only, as such, complete 400nm-700nm illumination, and, moreover, intensities of 400nm-700nm illumination that, in a sense, over-illuminate the observable universe—that is, illuminate it beyond what is necessary for complete illumination –, and ignoring the obvious implications for the organic and inorganic realms of the earth, we surely would be considerably neurologically and psychologically different. And if hypothetically we were perpetually subjected to such a state of illumination, we surely would be neurologically and psychologically devastated in an array of ways.

There clearly moreover is an explanation for why the macrolimit of the observable universe, itself, is not observable, measurable, and precisely conceivable, and why, instead, all that can be observed is black light in conditions outside of the surface illumination and atmospheric illumination of astronomical objects, and in various conditions of non-sunlight illumination on earth, such as starlight illumination, moon-reflection illumination, etc: The presence of a perpetual black background of the observable universe is (a) crucial in order to permit the aforementioned *(b)* of the micro external universe, and (b) it itself is immensely aesthetically, psychologically, and conceptually beneficial, as it, to list a few of a vast number of things that it accomplishes, provides a visual sense and conceptual sense of "limitlessness," allows for the content of the observable universe to be maximally highlighted, is pleasant to look at, is pleasant to know about, allows for the light of the astronomical light sources of the observable universe to be experienced as discrete light sources, etc.

As an aside, again regarding why the micro external universe and macrolimit of the observable universe are of black light: Realms of black light, including where observable matter is observed during 400nm-700nm illumination, are, moreover, visually, conceptually, and psychologically experienced as being of nothing, emptiness, vacuity, etc., and of rest, inactivity, etc. It is visually, conceptually, and psychologically beneficial to experience realms of the observable universe in this way, and for a vast array of reasons, such as to experience something that maximally contrasts with, or which is maximally opposing to, realms of observable matter and observable activity, and to know that such realms exist.

That black light has not been detected by particle collision experimentation, nor any other kind of experimentation, supports that it is something that is not of the observable universe, that it is formally different than anything of the observable universe, that it itself is unmeasurable and unobservable, that it is a fundamental property of the observable universe, and that only the expression of it, namely black illumination, can be observed via unaided vision and aided vision. Moreover, that black light (or black light illumination) has not been observed, nor conceived of, throughout the time of humanity to the present, supports that it is masterfully designed to be perpetually overlooked, and that the theory that would ensue from its observation is therefore intended to be maximally obviated.

115. Regarding the "hidden extra dimensions" of Randall, Susskind, et al., and setting aside the previously discussed problems thereof, and simply meaninglessly referring to them as "dimensions," they (that is, existents that have those dimensions) are conceived of as existing at an extraordinarily minute micro-depth from any location in the observable universe, and, as being the fundamental constituents of the entirety of the observable universe. Setting aside the problem of how, if they subsisted at a particular micro-depth, they would not be fundamental constituents of the observable universe, because they would not comprise the entirety of the observable universe, and setting aside the array of other problems that were discussed before, such as that the observable universe cannot be fundamentally composed of particles, and that particles, in any case, are not discrete existents, but rather, only relatively discrete existents, if they were at least considerably pervasive in the observable universe, it is the case that notwithstanding their extreme minute size, they (that is, their dimensionality) would be evident throughout much of the observable universe. That is, much of the entirety of the observable universe would be of their dimensions. As an aside, if there were infinitesimal existents of hidden extra dimensions that were fundamental existents or elementary existents of the observable universe, then their dimensionality would not be "hidden," and instead, highly observable: All complex existents of the observable universe would be entirely comprised of the elementary existents; and as such, all complex existents would be of the extra dimensions.

116. Since pure black light is of the 1-4 dimensional micro external universe, would the non-pure black light that is in the observable universe be 1-4 dimensional? It would be 1-3|5 dimensional, because it is expressed from the 1-3|5 dimensional microlimit of the observable universe. Notwithstanding, this is what surely additionally explains why black light has not yet been detected: In addition to what was mentioned previously, namely that black light is not of the observable universe, and that it, as such, is formally different than anything of the observable universe, its 1-3|5 spatial dimensionality is likely precisely why it has not been detected. Any 1-3|5 dimensional existent in the observable universe would not be physically accessible.

However, how then could non-pure black light be visually observable?: 1-3|5 dimensional existents would, in addition to being physically inaccessible, be sensorily inaccessible. Moreover, how then could black light accelerate the dispersion of the content of the observable universe?: 1-3|5 dimensional existents would, due to being formally different, and dimensionally different, not only be physically inaccessible, and sensorily inaccessible, but they would not interact with, nor affect, ordinary matter.

The visual observation of black light demonstrates that 1-3|5 dimensional existents can subsist in the observable universe, and that they could, as such, interact with the content of the observable universe. The problem of physically accessing them, or, alternatively, measuring their effects on what is physically accessible, is surely therefore due to our inability to directly access, or indirectly measure, 1-3|5 dimensional existents. However, and while this is not a measurement of even its general intricacy, black light has in fact been detected: It is what accelerates the dispersion of the content of the observable universe—that is, the dispersion of astronomical systems, such as galaxies—outward and away from one another. The movement of observable matter by black light is analogous to what is sought in the aforementioned experimentation: black light carries away the energy of ordinary matter. However, what is of course concurrently necessary for this "detection" of black light is the observation of it (that is, the observation of at least its expression, namely black light illumination), because otherwise there would be uncertainty about what is causing the accelerated dispersion. Aside: In the case of black matter, since only the expression of black light can be observed, nothing can be inferred about what even generally is adhering the intricacy of galaxies. That is, nothing can be inferred about what even generally black matter is.

117. Regarding the twenty-two principles of optical geometric physics in *(89)*, the following are modifications of some of the principles, replacements of some of the principles, and new principles.

(23) The microlimit of the observable universe is the three-dimensional surface of the observable universe. It is the fifth spatial dimension. Moreover, it is the entry to the fourth spatial dimension.

(24) The microlimit of the observable universe is a portion of the black matter of the micro external universe. Moreover, it expresses black light. The black light can be partly to completely supplanted for visual beings.

(25) The micro limit of the observable universe is the macrolimit of the micro external universe. Moreover, it is, from the perspective of the micro external universe, the entry to the 1-3|5 dimensional observable universe.

(26) The micro external universe is 1-4 dimensional.

(27) The observable universe is 1-3|5 dimensional.

(28) The macrolimit of the observable universe is a three-dimensional

boundary limit, and, moreover, the entry to the 1-4 dimensional macro external universe.

(29) The four dimensions of the micro external universe are inextricably combined; and the four dimensions of the observable universe are inextricably combined. That is, there are no existents that are only of one, two, or three of the dimensions. Moreover, dissociatively separating the dimensions for the purpose of constructing theory yields inaccurate theory.

(30) The black matter that is in the observable universe (namely, the micro limit of the observable universe) is 1-3|5 dimensional.

(31) The black light that is in the observable universe is 1-3|5 dimensional.

(32) Black matter emits black light; and black light, itself, expresses black light illumination.

(33) The black matter and black light of the micro external universe and macro external universe are 1-4 dimensional.

(34) The black matter of the micro external universe is the first cause of, and enduring cause of, and the mechanism of, the observable universe.

(35) Since the macro external universe three-dimensionally encompasses the observable universe, it, as is the case of the micro external universe, is 1-4 dimensional.

(36) Since the black matter of the micro external universe and macro external universe are of the same 1-4 dimensional realm, both are the first cause of, enduring cause of, and mechanism of, the observable universe.

(37) The observable universe can be pictorially reconceived of, and pictorially re-represented, as 1-3|5 dimensional, whereas, while the fourth dimension can be pictorially conceived of, and pictorially

represented, it, concurrent with the 1-3 dimensions, cannot be pictorially conceived of, nor pictorially represented. That is, the 1-4 dimensional external universe (that is, the micro external universe and macro external universe), and its existents, and dimensionality, cannot be accessed, nor accurately conceived of, nor accurately represented.

(38) The 1-4 dimensional external universe may of course consists of one or more additional spatial dimensions. However, if there are such dimensions, we are formally constrained from accessing existents of that dimensionality, and accurately conceiving of existents of that dimensionality, and accurately conceiving of that dimensionality itself.

118. As was discussed earlier, we are human by virtue of our formal inability to observe, access, enter, measure, and conceive of the micro external universe, macro external universe, microlimit of the observable universe, and macrolimit of the observable universe. If we had those abilities, we would be formally radically different than we are.

119. Nima Arkani-Hamed states the following about the possibility that the region of the observable universe that cannot currently be observed, and which may never be observed, due to how the light that travels from it will never reach earth, nor our technology that is located beyond earth, is "different" than our region of the universe.

> How are we supposed to deal with parts of the universe that we will never ever see again? More generally, this leads to the possibility that out there somewhere, even further away, there are parts of the universe that look very very different than our own universe, and that the particles that live [there], and the interactions that they have, might be different than the ones that live in our universe.[34]

---

34. Nima Arkani-Hamed on, "What are the Ultimate Questions of Nature?" *Closer To Truth,* 4:00-4:20, www.closertotruth.com/interviews/1709, 2016.

(For clarification, when Arkani-Hamed states the following (I add italics for emphasis), "… there are parts of *the universe* that look very very different than *our universe*," by, "our universe," he surely means our region of *the* universe. Also for clarification, while the entirety of the observable universe is observable, in the sense that all regions of it could be observed if there were observers who were located throughout it, there is, for humans, a practically non-observable region of it, due to the region being to far away for its light to reach us). What Arkani-Hamed more specifically means by the aforementioned region of the universe as possibly being "different" is that it is possibly formally different—that is, that it is possibly of formally different existents, and that the formally different existents are, moreover, of a formally different physics. This demonstrates that Arkani-Hamed believes that it is coherent to conceive of another region, or other regions, of the observable universe as being of formally different existents, and formally different physics, compared to our region of the observable universe. Andrei Linde likewise states the following:

> …. "the universe" can be divided into extremely large regions, which may have different laws of physics.[35]

While Linde's concept is an aspect of his theory of multiple universes, it notwithstanding is additional demonstration of the acceptance of the possibility that there are one or more regions of the universe where formally different existents subsist that are subject to formally different physics. (As an aside: For Arkani-Hamed, as was discussed, the region of the universe where this possibility might obtain is currently, and perhaps permanently, a non-observable region. Now, while the region is currently, and perhaps permanently, non-observable, it *could* be observed. That is, it is contingently non-observable. Moreover, Arkani-Hamed does not argue that it is of a different spatial dimensionality; nor

35. Robert Lawrence Kuhn, Space, "Confronting the Multiverse: What 'Infinite Universes' Would Mean," www.space.com/31465-is-our-universe-just-one-of-many-in-a-multiverse.html, December 23, 2015.

does he of course provide the basis or bases for thinking so. As such, the region is spatially contiguous with the observable universe.) For Linde, the regions ("infinite universes") are not contiguous, as they are separated by "disconnected space-time":

> We started calling it a 'multiverse', meaning the entire ensemble of innumerable regions of disconnected space-time ....[36]

What he means by "disconnected space-time" is the absence of space-time; and for him, the multiple universes are separated by these absences. However, the multiple universes are of course spatial, and subsist among one another spatially; and as such, the regions between them would be spatial as well. Moreover, Linde does not explicate what he means by the absence of space-time, nor does he consider how spatial multiple universes could subsist among one another if there were no spatial regions between them. Moreover, in his below pictorial representation of multiple universes, he does not represent what the absence of space-time is.

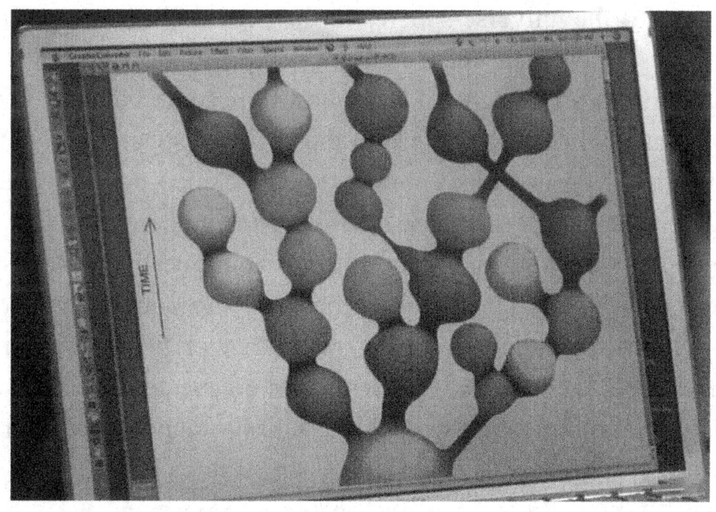

36. Ibid.

Regarding his pictorial representation, he states the following:

> .... this idea that the universe can be divided into the many regions .... we worked on showing it in pictures and showing it in computer simulations; and this is what is shown here in this picture. This is our universe looking like a huge fractal consisting of different extremely large parts. When you live in each of these parts, here or here, or here, for you it will look like almost infinite universe. .... But then see I've used here different colors; and I use here different colors to show that in each of these universes you may have slightly different laws of physics.[37]

Robert Lawrence Kuhn states the following about Linde's perspective:

> Linde portrays "universes" as painted balloons or bubbles on canvas, "squeezing off" from one another via eternal chaotic inflation. Each of his bubbles is a separate universe, each with different laws of physics.[38]

However, (a) the universe-spheres in the pictorialization are not separate from one another, but rather, are physically connected via tube-like structure; and the tube-like structures are quite large, as they are approximately 1/4 to 1/8 the size of the universe-spheres, (b) notwithstanding that the following is not a feature of Linde's pictorialization, if the universe-spheres "squeezed-off from one another," as Kuhn states of Linde's perspective, the squeezing-off would occur in 1-3|5 spatial dimensions; and the spatial region between the once conjoined universe-spheres would not suddenly disappear, or become absent: As was discussed, the universe-spheres are of course spatial, and, subsist among one another spatially; and as such, the regions between them

37. Andrei Linde on, "How Many Universes Exist?" 5:30-6:14, *Closer To Truth,* www.closertotruth.com/interviews/2606, December 2015.
38. Kuhn, Space, "Confronting the Multiverse: What 'Infinite Universes' Would Mean."

would be spatial as well. Moreover, Linde does not explicate what he means by the absence of space-time, nor does he consider how spatial universe-spheres could subsist among one another if there were no spatial regions between them. Moreover, in his above pictorial representation of universe-spheres, he does not represent what the absence of space-time is. (c) Linde's phrase, "disconnected space-time," by which he means, as was discussed, the absence of space-time, which of course is analogous to "the absence of space," is another instance of a meaningless antonymic linguistic alteration: He uses a generally coherent concept, and then linguistically antonymizes the names or phrases that are used to represent the concept; and there is no conceptual content for the antonymized names or phrases. Moreover, Linde engages in dissociation, as he dissociatively ignores that the regions between the universe-spheres are spatial. Moreover, he then, in place of what he dissociates from, uses the above meaningless antonymic phrase. That is, he replaces what he dissociates from with his meaningless antonymic phrase. (d) Linde states that he colors the different universe-spheres differently in order to represent that "in each of these universes you may have slightly different laws of physics." However, the use of different colors does not even minimally pictorially represent different laws of physics. Moreover, he does not even attempt to describe what the different laws of physics might be, and instead, uses meaningless language, and in this case, a meaningless type linguistic alteration, namely that various different types of physics may subsist. Aside, he conceives of what he states are different laws of physics as, themselves, existents: "in each of these universes you may have slightly different laws of physics." What should be stated is that there may be slightly different laws of physics that *explain* the intricacy of the physical of the universes. Laws are not existents that affect the physical, but rather, are concepts that explain the physical. (e) Regarding what precisely Linde means by "multiverse," he states the following, which was discussed in another context above:

> This is our universe looking like a huge fractal consisting of different extremely large parts. When you live in each of these parts, here or here, or here, for you it will look like almost infinite universe.[39]

For Linde, then, what determines whether a part is a separate universe is that it would be, for humans, sensorily, kinetically, and technologically experienced as a separate universe, or, as he states "almost [an] infinite universe." That is, that it would "look like" a separate universe to people within it. That is, that it, "for you," would "look like" a separate universe. Anthony Aguirre advances the same concept, and provides additional explanation:

> If I take this thing [he holds his palms about six inches apart] , and then a moment later maybe it is this big [he moves his palms about two feet apart], then a moment later maybe this big [he moves his palms about four feet apart]; so at each time, it has a finite size; but that's according to when I say "this moment," "a moment later," "a moment later." That's a definition of what a moment is, and what's existing at one time. Now I can take that same thing, and slice and dice it into space and time in a slightly different way, such that when you look at that, at every moment it looks like it is spatially infinite. The region you're allowed to look at is only the region inside this blob. You confine yourself inside this blob. But you can cut it up in such a way that inside that blob it looks like an infinite universe. . . . outside this blob there could be other stuff.[40]

Aguirre and the aforementioned Robert Lawrence Kuhn then have the following discussion-exchange during the above interview of Aguirre in which they discuss more then one "thing," "blob," etc.

---

39. Andrei Linde on, "How Many Universes Exist?" 5:30-6:14, *Closer To Truth,* www.closertotruth.com/interviews/2606, December 2015.
40. Anthony Aguirre on, "What Would an Infinite Universe Mean?" 1:27-2:31, *Closer To Truth,* www.closertotruth.com/interviews/1684, November 2014.

Aguirre: .... in the external description, in which they look finite ....(3:26-3:29)

Kuhn: [Restating Aguirre's position] .... looked at from outside, they both look finite. (3:43-3:46)

Aguirre: [Agreeing] They both look finite. (3:47)

Kuhn: But inside? (3:48)

Aguirre: [Answering] They both look infinite. (3:49)[41]

As is the case for Linde, what is relevant for Aguirre is where the observer is located: If the observer is located within a region, or blob, or part, etc., and it "looks" like a separate universe for the observer, then it is. That is, for Aguirre, "You confine yourself inside this blob. .... inside that blob it looks like an infinite universe," and for Linde, "When you live in each of these parts, here or here, or here, for you it will look like almost infinite universe." Aguirre does state, though, that for an observer who is outside of the region, blob, part, etc., the region, blob, part, etc., "look[s] finite." However, he does not employ his aforementioned observational logic by stating that since it looks finite from the outside perspective, it is finite.

Both Linde and Aguirre, however, ignore the core issue, namely whether the region, blob, etc., in question is objectively a separate universe; and instead they advance a theory of relativity according to which the region is finite from any perspective that is external to the region, and infinite from any perspective that is inside the region.

As is implied by my previous discussions about the issue of whether what appear to be discrete existents of the observable universe have macrolimits, in order for there to be discrete universes, just as would be necessary in order for there to be discrete existents, there would have to be external universe between the universes. (In order for there to be discrete existents, there would have to be external universe between

41. Ibid.

the existents). I demonstrated previously that since there is no external universe between what appear to be discrete existents of the observable universe, there are no discrete existents, and that, as such, the existents are relatively discrete existents, or relative existents. Not only is the same the case of what is theorized to be a multiplicity of discrete universes, the objection is even more severe, since a multiplicity of what appears to be discrete universes has never been observed, unlike the multiplicity of what appears to be discrete existents in the observable universe.

Moreover, if hypothetically there were discrete universe-spheres, or discrete universe-realms, and, as such, external universe between them, this would not be discernible, since, as was discussed previously, it would be formally impossible for the beings of the observable universes to exceed the macrolimits of the observable universes. All that could be done is to hypothesize that may be discrete universe-realms.

Notwithstanding the above various discussions, an issue that is not attended to by Linde nor Aguirre is that even if there were discrete universe-realms, the totality of the universe-realms could nevertheless be considered to be one universe. That is, the totality of the universe-realms could be considered to be sub-universes of one universe, and that the sub-universes are separated by external universe, which, itself, is a facet of the one universe.

The general concept of multiverse is, itself, dissociative, because it dissociates from what was discussed above about the spatial relation, or external universe relation, between universe-realms. It is moreover meaningless because an aspect of it is the aforementioned meaningless antonymic linguistic alteration, namely that space-time disconnects, or discontinues, or is absent, between the universe-realms. While such phrases are psychologically puissant, and the second stage of the psychological approach to the issue (the first stage is the aforementioned psychologically motivated conceptual dissociation; and the second stage is the ensuing substitution of what was dissociated from with the meaningless antonymic linguistic alteration), the approach is psychological, rather than scientific, and even scientific-philosophical, and even predominately philosophical. Aside, the word "breakdown," and phrases thereof, such as "space and time breakdown," "the laws of physics breakdown," etc., are used by many scientists and scientist-philosophers; and the aforementioned two stage psychological process is what is involved, namely, firstly, psychologically motivated conceptual dissociation, and secondly, and ensuingly (that is, the dissociation gives rise to the possibility of it), the production of conceptually meaningless, yet psychologically puissant, linguistic alterations.

120. Nima Arkani-Hamed states the following in which he argues that mathematics that utilize "space-time locality" overlook "wonderful mathematical structures hidden in quantum field theory and string theory," and that "a new way of thinking about quantum field theories" may result in the "remarkable hidden structures" being "made manifest." He further argues that "by removing space-time from its primary place in our description of standard physics" we may be able to ascend to "the next theory," namely "where space-time finally ceases to exist." He believes that jettisoning space-time mathematics may result in the ascension to a more evolved level of mathematics in which the mathematics that is currently hidden in quantum field theory and string theory may be uncovered. That is, mathematics in which "space-time finally ceases to exist" is where the "hidden" "wonderful mathematical structures" are.

> Thus, while we may not have experimental data to tell us about physics near the Planck scale, we do have an ocean of "theoretical data" in the wonderful mathematical structures hidden in quantum field theory and string theory. These structures beg for a deeper explanation. The standard formulation of field theory hides these amazing features as a direct consequence of its deference to space-time locality. There must be a new way of thinking about quantum field theories, in which space-time locality is not the star of the show and these remarkable hidden structures are made manifest. .... by removing space-time from

> its primary place in our description of standard physics, we may be in a better position to make the leap to the next theory, where space-time finally ceases to exist.[42]

Arkani-Hamed's basis for the above is the following, namely that "near [physically near] the "big-bang," "our descriptions of physics break down along with the notion of time itself."

> Looking back in time, we eventually encounter Planckian space-time curvatures near the "big bang," where all our descriptions of physics break down along with the notion of time itself.[43]

Natalie Wolchover further explains some of the above about Arkani-Hamed's perspective in the following statement in which she states that "in certain situations involving gravity," both "locality" (namely, "particles can interact only from adjoining positions in space and time"), and "unitarity" (namely, "the probabilities of all possible outcomes of a quantum mechanical interaction must add up to one"), "break down," "suggesting neither is a fundamental aspect of nature."

> Locality is the notion that particles can interact only from adjoining positions in space and time. And unitarity holds that the probabilities of all possible outcomes of a quantum mechanical interaction must add up to one. The concepts are the central pillars of quantum field theory in its original form, but in certain situations involving gravity, both break down, suggesting neither is a fundamental aspect of nature.[44]

---

42. Nima Arkani-Hamed, "The Future of Fundamental Physics," *Daedalus,* Journal of the American Academy of Arts & Sciences, p. 66, www.amacad.org/publication/future-fundamental-physics, www.sns.ias.edu/ckfinder/userfiles/files/daed_a_00161(1).pdf, Summer 2012, p. 66.
43. Ibid., p. 59.
44. Natalie Wolchover, "A Jewel at the Heart of Quantum Physics," *Quanta Magazine,* https://www.quantamagazine.org/physicists-discover-geometry-underlying-particle-physics-20130917/, September 17, 2013.

Wolchover further explains the discussion with the following statement in which she states that "Einstein's theory [Einstein's general theory of relativity] and the space-time concept break down inside black holes and at the moment of the big bang." She further explains the discussion by indicating that there may be a "more abstract or unfamiliar" "description of reality" that "can have greater explanatory power" than that of "space-time," which she states may be a "translation" of the "more abstract or unfamiliar" "description of reality."

> Today, various puzzles and paradoxes point to the need to reformulate the theories of modern physics in a new mathematical language. Many physicists feel trapped. They have a hunch that they need to transcend the notion that objects move and interact in space and time. Einstein's general theory of relativity beautifully weaves space and time together into a four-dimensional fabric, known as space-time, and equates gravity with warps in that fabric. But Einstein's theory and the space-time concept break down inside black holes and at the moment of the big bang. Space-time, in other words, may be a translation of some other description of reality that, though more abstract or unfamiliar, can have greater explanatory power.[45]

For Arkani-Hamed, space-time physics and mathematics, including "the notion of time itself," "break down" "near the "big bang""; and for Wolchover, via her perspective about Arkani-Hamed's perspective, "Einstein's theory and the space-time concept break down inside black holes and at the moment of the big bang." However: (a) Physics, mathematics, concepts, theory, descriptions, laws, principles, etc., do not "break down" in any physical context. Such things are of the mind (or mental); and as such, they are are not affected by the physical contexts in which they arise, and they do not affect the physical contexts in

45. Natalie Wolchover, "A Different Kind of Theory of Everything," *The New Yorker*, www.newyorker.com/science/elements/a-different-kind-of-theory-of-everything, February 19, 2019.

which they arise. Believing that they do entails misconceiving of them as physical existents, and as physical existents which, moreover, have a causal relation with the physical contexts in which they arise. The misconception firstly involves dissociating from what the existents are, namely mental existents (or features of the mind), and then secondly involves delusively reconceiving of them as physical existents. The dissociation provides for the possibility of the delusive reconception. That is, the dissociation removes the constraint on delusive reconceptualization. As such, Einstein's theory and the space-time concept do not break down inside black holes and at the moment of the big-bang, and space-time physics and mathematics, including the notion of time itself, do not break down near the big-bang. Rather, the various mental activity (the physics, mathematics, concepts, etc) are simply inapplicable for accurately describing and explaining the intricacy of black holes, and for providing an accurate theory of the cause of the big-bang. (b) Elsewhere in the discussions of Arkani-Hamed et al., space, time, space-time, etc., themselves, are conceived of as breaking down in certain contexts, meaning that they do not exist in those contexts. Again, Arkani-Hamed states the following:

> .... by removing space-time from its primary place in our description of standard physics, we may be in a better position to make the leap to the next theory, where space-time finally ceases to exist.

He firstly discusses removing space-time physics from standard physics, and then discusses that in the physics that ensues, space-time, itself, will not exist. Of course he does not mean, in stating, "where space-time finally ceases to exist," that only the concept of space-time finally ceases to exist. Likewise, where he states, "by removing space-time from its primary place in our description of standard physics," he of course does not mean removing space-time itself, but rather, space-time physics. Wolchover, in her below statement about this discussion, demonstrates the conceptual process that occurs in Arkani-Hamed:

> Locality is the notion that particles can interact only from adjoining positions in space and time. And unitarity holds that the probabilities of all possible outcomes of a quantum mechanical interaction must add up to one. The concepts are the central pillars of quantum field theory in its original form, but in certain situations involving gravity, both break down, suggesting neither is a fundamental aspect of nature.

For Wolchover, and Arkani-Hamed, since the concepts are inapplicable for theory that contends with describing and explaining certain situations of gravity, what the concepts are of, such as, for example, space-time, either do not exist, or are not fundamental aspects of nature. That is, since the concept of space-time does not describe and explain the causal condition or causal conditions of the big-bang, and the intricacy of black holes, it is the case, for Arkani-Hamed, that not only is space-time not a fundamental aspect of nature, it does not exist. Aside, it is unclear what Wolchover means by "neither is a fundamental aspect of nature," as she does not explain the difference between conceiving of space-time as a fundamental aspect of nature, and conceiving of it as not a fundamental aspect of nature. Moreover, since space-time is always conceived of as a fundamental aspect of nature, it appears that her belief that it can be reconceived of as not a fundamental aspect of nature is inaccurate. Moreover, Arkani-Hamed does not discuss that in the evolved physics that he discusses, space-time is non-fundamental. Rather, he discusses that it does not exist.

Arkani-Hamed illustrates his above 2012 discussion with the following 2014 discussion about a geometric object that he and Jaroslav Trnka constructed, which he refers to as the "amplituhedron." As an aside, in his 2012 discussion (namely, the aforementioned article, *The Future of Fundamental Physics*) he, as was discussed, discussed the ceasing-to-exist of space-time, and the ensuing discontinuation of the use of space-time physics. In his 2014 discussion, he advances a theory according to which not only is space-time not a fundamental aspect of the universe, the concepts of space and time are constructs that are inaccurate—that is, they do not represent any aspect of the universe. While the constructs are useful for most people in constructing an understanding of the universe, and in having understandable sensory experiences of the universe, and for having socially intelligible discussions about matters of simple practical experiences, to complex observations and theories, they are merely contrived constructs that do not represent the way that the universe actually is. Arkani-Hamed states the following about the amplituhedron; and of course, the "locality and unitarity" that he mentions were discussed above.

> .... where locality and unitarity do not play a central role but are derived consequences from a different starting point. In this note we provide such an understanding .... which we identify as "the volume" of a new mathematical object—the Amplituhedron—generalizing the positive Grassmannian. Locality and unitarity emerge hand-in-hand from positive geometry.[46]

Wolchover provides the following elaboration:

> Physicists have discovered a jewel-like geometric object that dramatically simplifies calculations of particle interactions and challenges the notion that space and time are fundamental components of reality.....
>
> The amplituhedron is not built out of space-time and probabilities; these properties merely arise as consequences of the jewel's geometry. The usual picture of space and time, and particles moving around in them, is a construct.....
>
> Beyond making calculations easier or possibly leading the way to quantum gravity, the discovery of the amplituhedron could cause an even more profound shift, Arkani-Hamed said. That is, giving up space and time as fundamental constituents

46. Nima Arkani-Hamed and Jaroslav Trnka, "The Amplituhedron," *Journal of High Energy Physics,* https://arxiv.org/pdf/1312.2007.pdf, abstract, Dec. 6, 2013.

> of nature and figuring out how the Big Bang and cosmological evolution of the universe arose out of pure geometry. "In a sense, we would see that change arises from the structure of the object," he said. "But it's not from the object changing. The object is basically timeless."[47]

(a) In stating that Arkani-Hamed and Trnka have "discovered a jewel-like geometric object that … challenges the notion that space and time are fundamental components of reality," Wolchover apparently believes that they discovered a physical object. Moreover, in the following additional statement of hers, there is further indication that she believes that they discovered a physical object: "The amplituhedron is not built out of space-time and probabilities." Here she indicates that while the amplituhedron "is not built out of space-time," it is nevertheless built out of something. Arkani-Hamed, also, at times, apparently conceives of the amplituhedron as a physical object: He states that the object has a "structure," and that it is "basically timeless." As such, both Wolchover and Arkani-Hamed describe and discuss the amplituhedron as if it is a kind of physical object, or existent, in the universe. However, Arkani-Hamed, at other times, describes it as a "mathematical object."[48] Mathematical objects, though, do not have physical structure, but rather, conceptual structure; and the only "physical structure" that they may have is conferred to them via the graphical representations of them; but graphical representations, while graphical, and physical in the sense of being graphical, are not physical objects, or existents, of the universe. Arkani-Hamed also describes the amplituhedron as "basically timeless"; and aside from how, as was discussed earlier, the concept of timelessness is dissociative, and the language "timeless," "timelessness," etc., is linguistically antonymic,

47. Natalie Wolchover, "A Jewel at the Heart of Quantum Physics," *Quanta Magazine*, https://www.quantamagazine.org/physicists-discover-geometry-underlying-particle-physics-20130917/, September 17, 2013.

48. Nima Arkani-Hamed and Jaroslav Trnka, "The Amplituhedron," *Journal of High Energy Physics,* abstract, p. 3., https://arxiv.org/pdf/1312.2007.pdf,

when something's temporality is analyzed, and thereafter judged to be timeless, or "basically timeless," the something whose temporality is analyzed is not a conceptual object, nor a graphical object, because such things are not considered to have temporal natures of their own: conceptual objects subsist in the conceivers for as long as the objects are conceived and remembered; and graphical objects subsist on, or in, the various physical mediums where they are physically rendered, for as long as the renderings remain, or are stored. That is, their temporality is entirely dependent on their creators, and the mediums on which, or in which, they are rendered by their creators, and possibly stored. As an aside, regarding Arkani-Hamed's conception of the amplituhedron as "basically timeless," this is further indication, and perhaps the clearest indication, that he conceives of the amplituhedron as a physical object, and not a mathematical object: In what sense would a mathematical object be "basically timeless"? All mathematical objects have the same temporal nature: It is not the case that some mathematical objects are basically timeless (or essentially timeless, or predominately timeless), while others are truly timeless. Moreover, as discussed above, mathematical objects do not have temporality of their own. As such, Arkani-Hamed is referring to a physical existent in the observable universe that does not depend on anything for its existence, nor subsistence; and this would explain the nature of Wolchover's discussion. (b) The following is an additional and even clearer demonstration that Arkani-Hamed, Trnka, and Wolchover conceive of the amplituhedron as a physical object; and the following is, moreover, the first demonstration that their conception of the amplituhedron, as well as their conception of its causal nature, are severely dissociative.

> The amplituhedron is not built out of space-time and probabilities; these properties merely arise as consequences of the jewel's geometry.
>
> Locality and unitarity emerge hand-in-hand from positive geometry.

> That is, giving up space and time as fundamental constituents of nature and figuring out how the Big Bang and cosmological evolution of the universe arose out of pure geometry. "In a sense, we would see that changc arises from the structure of the object," he said. "But it's not from the object changing …."

For Arkani-Hamed, Trnka, and Wolchover, the amplituhedron's geometry is what caused space and time (that is, it is what caused "the Big Bang and cosmological evolution of the universe" to "emerge." The "pure geometry" of the amplituhedron—that is, "the structure of the object"—is the cause; and since this pure geometry is the cause, it is referred to as "positive geometry." And for clarification about how the object's pure geometry is causal, Arkani-Hamed, et al., adduce that the causation does not occur via the object changing in nature. However, Arkani-Hamed, et al., dissociate an object's geometrical nature from the object: Since an object's geometry is an inextricable aspect of the object, when it is dissociated from, the object would cease to exist. That is, an object's geometry cannot exist separately from the object. Likewise, the object cannot exist separately from its geometry.

To elaborate on my above discussion: (a) Any object, by virtue of it being an object, has a geometry, which of course could be an ever-changing geometry. (b) Objects, by virtue of being objects, are observable: If there is nothing to observe, then there are no objects. (c) An object without a geometry has never been observed; and since, moreover, an object without a geometry cannot be even minimally coherently conceived of, there is, therefore, no such object. (d) Mathematical language that expresses that a particular object's geometry is separate from the object is language of what I will refer to as "meaningless separative linguistic alteration." Simply because a dissociative non-mathematical concept can be expressed mathematically does not entail that the mathematical expression is even minimally coherent. Mathematics should always be founded upon non-mathematical conceptual coherence, rather than the variety of meaningless linguistic alterations, because otherwise it itself will be of the nature of one or more of the variations of meaningless linguistic alteration.

In addition to above discussions, and setting aside what was revealed about the amplituhedron as an object, the amplituhedron as a non-mathematical concept, and the amplituhedron as a mathematical concept, setting aside how, as was discussed earlier, the concept of timelessness is dissociative, and the language "timeless," "timelessness," etc., is linguistically antonymic, I would like to incoherently allow that the basically timeless pure positive geometry that Arkani-Hamed, et al., believe subsists in the universe is the case; and what I am allowing is a mere semblance of some kind of thing. For Arkani-Hamed, it is an existent, or object; and as such, the problem of what caused it is not successfully evaded, nor is the problem of how it caused the universe. Arkani-Hamed, et al., attempt to postulate that there is a kind of existent that exists outside of time (or what I previously discussed is "duration of existence"), due to it being only of geometry; yet all existents, of any kind, have duration-of-existence; and moreover, dissociating an inextricable aspect of an object from the object does not succeed at making it possible that that aspect of the object is timeless, and causative of the universe: The object exists, and as such has duration-of-existence; and the object exists in the universe, which gives rise to the problems of what caused it, and how it causes other things.

Aside from the above discussions, Arkani-Hamed, et al., do not provide an even minimal discussion about how an object can be composed of only its geometry (that is, how an object can be only geometric in nature, and not, for example, have a content), and how "the volume" of an object is its geometry. Moreover, as for whether Arkani-Hamed and Trnka provide a graphical representation of the amplituhedron, they do provide the following representation, and qualify it by stating the following:

> Just to have a picture, below we sketch a 3-dimensional face of the 4 dimensional amplituhedron for n = 8, which turns out

to be the space $Y = c_1Z_1 + \ldots c_7Z_7$ for Z_a positive external data in P^3.[49]

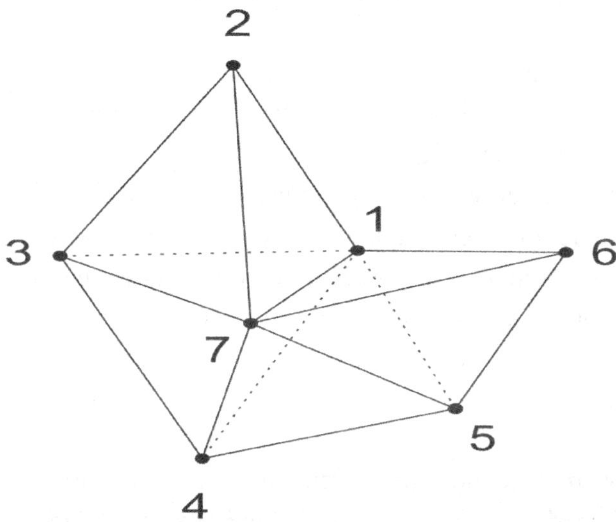

Aside, regarding the dimensionality of the amplituhedron, Wolchover explains the following:

> The details of a particular scattering process dictate the dimensionality and facets of the corresponding amplituhedron.[50]

And in this sense, the amplituhedron can be thought of as being n-dimensional. Also aside, Anil Ananthaswamy provides the following explanations about the dimensionality of the amplituhedron, and the general nature of the amplituhedron:

> .... physicists are discovering hints of deeper geometric truths. If glimpses of exquisite mathematical structures that exist in

49. Ibid., p. 8.

50. Natalie Wolchover, "A Jewel at the Heart of Quantum Physics," *Quanta Magazine*, https://www.quantamagazine.org/physicists-discover-geometry-underlying-particle-physics-20130917/, September 17, 2013.

> dimensions beyond the familiar few can be substantiated, they would seem to point the way to a better understanding not just of how particles interact, but of the nature of reality itself.[51]
>
> It's best to think of the amplituhedron not as a real object but as an abstraction. It's a mathematical structure that gives us an elegant way to encode the calculations that tell us how likely a particle interaction is to play out in a certain way. The details of the interaction, meaning the number and properties of the particles involved and the forces involved, dictate the dimensions and facets of the corresponding amplituhedron—and that contains the answer. So there are actually many amplituhedra, one for each possible way in which a set of particles may interact.[52]

Ananthaswamy secondly discusses that it is inaccurate to conceive of the amplituhedron "as a real object," because it is "an abstraction," namely "a mathematical structure." However, he firstly discusses the following:

> .... physicists are discovering hints of deeper geometric truths. If glimpses of exquisite mathematical structures that exist in dimensions beyond the familiar few can be substantiated ....

Here he demonstrates that he believes that at least some mathematical structures are physical objects (that is, real objects, or non-conceptual objects), and that they are not, as such, abstract objects (that is, conceptual objects): In implying that the amplituhedron (which is the subject of his article) exists in dimensions, he is demonstrating that he believes that it is a physical object: There is no coherent sense in stating that particular mathematical objects (or mathematical structures) "exist," and

---

51. Anil Ananthaswamy, "The New Shape of Reality," *New Scientist,* Issue 1377, p.1, https://trnka.faculty.ucdavis.edu/wp-content/uploads/sites/322/2016/02/new-scientist.pdf, July 29, 2017.
52. Ibid., p. 2.

exist "in dimensions," and exist in "dimensions beyond the familiar few." As Ananthaswamy himself states:

> It's best to think of the amplituhedron not as a real object but as an abstraction. It's a mathematical structure that gives us an elegant way to encode the calculations that tell us how likely a particle interaction is to play out in a certain way.

That is, it is a purely conceptual object, which as such does not subsist as matter, and which as such does not interact with matter. It is confined to the mind, and may be generally physically represented via graphics, and via constructed models.

Clearly, then, the above people should state that the amplituhedron is speculated to exist, and that if it did exist, it would have the aforementioned nature. Moreover, the above people should state that in the meantime, the concept of the amplituhedron is useful in physics in various ways.

Now returning to the issue of the dimensionality of the amplituhedron, and setting aside the aforementioned problems about the concept of the amplituhedron, first what is meant by an object that does not subsist in space (or "space-time")? (That is, what is meant by an object that does not subsist within the microlimit of the observable universe)? As an aside: Notwithstanding that Arkani-Hamed, et al., do not state the following, it is clear that they conceive of the amplituhedron as subsisting outside of the observable universe. That is, they conceive of it as existing prior to the observable universe, and as being the cause of the observable universe. Now, as is shown by the above approximate graphical representation of it, it is an object that is, by virtue of being an object, at least, as was discussed earlier, three-dimensional; (and any extra dimensionality that it is purported to have is, as I discussed earlier, of meaningless additive linguistic alteration). It is interesting that Arkani-Hamed, et al., do not argue that the amplituhedron would, in addition to not subsisting in space, subsist outside of spatial dimensionality. How can an existent that does not subsist in space have spatial dimensions? The Institute for Advanced Study intimates that the amplituhedron (the multitude of them) are,

> .... the building blocks out of which spacetime emerges .... [53]

As such, they, like Arkani-Hamed, et al., conceive of the amplituhedron as not existing in space-time, and as causing space-time. Again, how can something that does not exist in space-time have spatial dimensions? Arkani-Hamed, et al., do not even attempt to provide an explanation. Moreover, what they do attempt is to furtively obviate this conspicuous problem by dissociating an existent's geometric nature from the existent, and by arguing that the non-existent geometry, or objectless geometry, is what exists outside of space-time, and is what causes space-time. However, as was discussed, there is no such thing as a geometry that is independent from any existent. Moreover, Arkani-Hamed, et al., are not able to even minimally graphically represent such a geometry. Moreover, they oscillate between characterizing the amplituhedron as a real object (a physical object) and as a mathematical object (an abstract object, or conceptual object); and their characterizations are at times explicit, and at other times implicit (that is, they discuss the amplituhedron as if it is a physical object at times, and as if it is a conceptual object at other times). Moreover, they oscillate between characterizing mathematical objects as physical objects and conceptual objects; and as such, this is a sub-oscillation within the aforementioned primary oscillation.

As an aside, and again, Wolchover explains an aspect of the discussion of Arkani-Hamed by indicating that there may be a "more abstract or unfamiliar" "description of reality" that "can have greater explanatory power" than that of "space-time," which she states may be a "translation" of the "more abstract or unfamiliar" "description of reality." She conceives of his concept of causal pure geometry as

53. The Institute for Advanced Study, "Nima Arkani-Hamed on the Amplituhedron," www.ias.edu/ideas/nima-arkani-hamed-amplituhedron, 2015.

being more abstract than the other concepts that attempt to describe the causal structure of the universe; and she moreover thinks that the greater explanatory power of his concept is due to its higher level of abstraction. However, his concepts of pure geometry, and causal pure geometry, are not more abstract: They instead are severely dissociative; and this may give the appearance that they are more abstract than the preexisting concepts, and, or, more accurate and more puissant than the preexisting concepts: As was discussed earlier, furtively dissociative concepts, and the supervening meaningless linguistic alterations, and that both are significantly enigmatic, can be immensely psychologically compelling; moreover, due to the pervasive failure to recognize conceptual dissociation in physics and mathematics—and conceptual dissociation is also prevalent in the foundation of ethics, and in meta-ethics, and other disciplines—the dissociation is typically experienced as ingenuous abstraction, and the abstrusity is typically experienced as the challenge of ingenuous abstraction.

Also aside, it is interesting that Ananthaswamy states the following:

> It's best to think of the amplituhedron not as a real object but as an abstraction.

It should be self-evident whether it is an object or concept; and that he provides such a recommendation is further demonstration of the aforementioned oscillation of Arkani-Hamed, et al., and even of a corollary effect of severe conceptual dissociation, as well as of the use of meaningless linguistic alterations to explain the dissociative concept, namely conceptual dissonance. (The dissonance, though, is seemingly never found to be a basis to analyze the formal nature of the concepts and supervening language; and again, there is a pervasive failure to recognize conceptual dissociation in physics, mathematics, etc. Moreover, the dissonance is actually typically experienced as an aspect of what is already experienced to be the challenging abstrusity of the concepts and language; and as such, it typically augments the psychological puissance).

The following is an elaboration of a fundamental facet of the perspective of Arkani-Hamed and Trnka, namely that the "geometry" of the amplituhedron is "timeless," "atemporal," "replaces the notion of time," of "timeless math," "does not exist in space-time," "is not rooted in a world where a particle starts in one place and time," is "outside" of "space-time," and is not "woven of this fabric [the fabric of space-time]."

Where Time Comes From

> The amplituhedron reconceptualized colliding particles—ostensibly temporal events—in terms of timeless geometry. When it was discovered in 2013, many physicists saw yet another reason to think that time must be emergent—a variable that we perceive and that appears in our coarse-grained description of nature, but which is not written into the ultimate laws of reality. At the top of the list of reasons for that hunch is the Big Bang. The Big Bang was when time as we know it sprang forth. Truly understanding that initial moment would seem to require an atemporal perspective. "If there's anything that asks us to come up with something that replaces the notion of time, it's these questions about cosmology," Arkani-Hamed said. Thus, physicists seek timeless math that generates what looks like a universe evolving in time. The recent research offers glimpses of how that might work.[54]
>
> …. the amplituhedron is not rooted in a world where a particle starts in one place and time before moving to the next location and moment. That is, the shape does not exist in space-time ….
>
> "We've known for decades that space-time is doomed," says

54. Natalie Wolchover, "Cosmic Triangles Open a Window to the Origin of Time," *Quanta Magazine,* https://www.quantamagazine.org/the-origin-of-time-bootstrapped-from-fundamental-symmetries-20191029/, October 29, 2019.

> Arkani-Hamed. "We know it is not there in the next version of physics." Though the collisions described by the amplituhedron still occur in space-time, the object itself is outside it, providing a possible way to imagine a world not woven of this fabric.[55]

Similarly, Stephen Hawking argues that time began at the event of the Big Bang, due to how "all of the laws of science" "break down" in the condition that caused the event, namely the "infinitesimally small and infinitely dense" "big bang singularity," where "the density of the universe and the curvature of space-time would have been infinite."

> Hubble's observations suggested that there was a time, called the big bang, when the universe was infinitesimally small and infinitely dense. Under such conditions all the laws of science, and therefore all ability to predict the future, would break down. If there were events earlier than this time, then they could not affect what happens at the present time. Their existence can be ignored because it would have no observational consequences. One may say that time had a beginning at the big bang, in the sense that earlier times simply would not be defined.[56]

> All of the Friedmann solutions have the feature that at some time in the past (between ten and twenty thousand million years ago) the distance between neighboring galaxies must have been zero. At that time, which we call the big bang, the density of the universe and the curvature of space-time would have been infinite. Because mathematics cannot really handle infinite numbers, this means that the general theory of relativity (on which Friedmann's solutions are based) predicts that there is a point in the universe where the theory itself breaks down. Such

55. Shannon Palus, "Amplituhedron May Shape the Future of Physics," *Discover Magazine,* www.discovermagazine.com/the-sciences/amplituhedron-may-shape-the-future-of-physics, January 7, 2014.
56. Stephen Hawking, *A Brief History of Time*, Bantam Books, 1998, p. 9.

> a point is an example of what mathematicians call a singularity. In fact, all our theories of science are formulated on the assumption that space-time is smooth and nearly flat, so they break down at the big bang singularity, where the curvature of space-time is infinite. This means that even if there were events before the big bang, one could not use them to determine what would happen afterward, because predictability would break down at the big bang.
>
> Correspondingly, if, as is the case, we know only what has happened since the big bang, we could not determine what happened beforehand. As far as we are concerned, events before the big bang can have no consequences, so they should not form part of a scientific model of the universe. We should therefore cut them out of the model and say that time had a beginning at the big bang.[57]

Perhaps Arkani-Hamed and Trnka would argue that only the amplituhedron existed prior to the Big Bang, as that is consistent with their argument that it exists outside of time, and that it, moreover, caused time, via its pure geometry. However, and conceiving of time as I argued for previously in *(19)* (namely, as the duration of existence of existents, as measured by regularly proceeding accretional mechanisms), the concept of a timeless existent is dissociative, as discussed above. Moreover, Arkani-Hamed and Trnka do not provide any arguments that even attempt to demonstrate that the amplituhedron is of this nature. Moreover, any existent that is causal of other existents, as is argued of the amplituhedron, is (a) an existent, and (b) temporal. Arkani-Hamed and Trnka engage in maximal and multifaceted dissociation in order to try to persuade that there is a kind of existent that is only of geometry (and as such, not truly an existent, or, not an existent at all), and, that it does not subsist in space nor time (that is, that it does not subsist in the universe).

57. Stephen Hawking, *A Brief History of Time*, Bantam Books, 1998, p. 49.

As is demonstrated by Arkani-Hamed's above meta-discussions, and the above elaborations of his meta-discussions, it is clear that he finds that he and others have not been able to successfully address the issues of physics that they study; and it is moreover clear that he engaged in the aforementioned unprecedented dissociation in order to try to obviate the array of problems that he and others have not been able to address: He dissociates a facet of an object from itself, and believes that the facet exists independently; he dissociates an object from the universe; he dissociates an object from temporality (that is, from time, or duration-of-existence as measured by mechanisms); and he nevertheless argues that the object causes time and space. By doing this, he evades an array of fundamental problems, and believes that he succeeds at providing an account of the origin of the universe. If he is asked what caused the amplituhedron, where it was located, when it emerged, and what its nature is, he would state that because it is not in the universe, nor temporal, and is only of geometry, such questions are therefore irrelevant.

Hawking conceives of the aforementioned singularity in a similar way. He conceives of the Big Bang as the beginning of time, and that, as such, the singularity from which the Big Bang emerged was outside of time. Interestingly, he does not argue that the emergence of the singularity is when time began. But, as is the case with Arkani-Hamed regarding the amplituhedron, by doing this he obviates an array of issues. Also interestingly, Hawking refers to the Big Bang as "a time," rather than an event. Again:

> Hubble's observations suggested that there was a time, called the big bang, when the universe was infinitesimally small and infinitely dense.

However, events are not times. Events occur at particular times, but are not themselves times. Moreover, he clearly should have stated the following:

> "Hubble's observations suggested that there was a time, namely when the singularity existed, when the universe was infinitesimally small and infinitely dense."

The Big Bang, itself, was not infinitesimally small and infinitely dense. Rather, the existent from which it emerged was. Aside, he alternatively could have stated the following:

> "Hubble's observations suggested that there was a time, namely before the Big Bang, when the universe was infinitesimally small and infinitely dense."

He clearly obviated stating both of the above statements because they express that time existed before the Big Bang. By conceiving of the Big Bang as "a time" ("there was a time, called the big bang"), he evades the following issues: when the singularity emerged; what caused it; and how long it subsisted before the Big Bang. He moreover argues that since the laws of physics did not apply prior to the Big Bang, time did not exist prior to the Big Bang. Aside, he demonstrates that he believes that the presence of various practical problems warrant ignoring "[i]f there were events earlier than this time ….":

> If there were events earlier than this time, then they could not affect what happens at the present time. Their existence can be ignored because it would have no observational consequences. One may say that time had a beginning at the big bang, in the sense that earlier times simply would not be defined.

For Hawking, due to how the singularity cannot be defined, and due to how the issues of when it emerged, what caused it, and how long it subsisted prior to the Big Bang, cannot be addressed via the laws of physics, the issue of "[i]f there were events earlier than this time" can be ignored. However, practical problems should not warrant the rejection of outstanding issues as being irrelevant. Moreover, he could have stated that there are many practical problems, and that a formally different physics, or additional physics in the current realm of physics, is necessary in order to address the issues, and that due to the absence of such physics, such issues currently cannot be addressed.

As an aside, Arkani-Hamed and Trnka do not address the issue of the current existential status of the amplituhedron. That is, they do not address whether they believe that it still exists, or whether it once existed, and that when it did, it caused time and the observable universe to emerge, and that after its causal period, it discontinued existing. Regarding whether it still exists, they do not address the issue of what effects it has now, given that it already caused time and the universe. For example, does it subsist outside of the universe in an incipient state, and as such have no effects; and if so, how does its nature change from being that of "positive geometry" to non-causal and incipient? Or, does it perpetually cause time and the universe, such that without it, the universe would, for example, immediately disintegrate and then cease to exist, or gradually disintegrate and then cease to exist?

Also aside, Arkani-Hamed and Trnka do not address the issue of how long the amplituhedron existed before it became causal of the universe. The causal dynamic between the amplituhedron and the universe is a temporal event; and based on the causal dynamic, the universe was caused. As such, and given that the amplituhedron is "timeless" (or eternal), it existed prior to causing the universe. The following issues therefore arise: How long did the amplituhedron exist prior to causing the universe; what were its effects during that time; and why did it, at a particular time, cause the universe?

Also aside, Arkani-Hamed and Trnka do not even attempt to explain what it is for something to exist outside of the universe, and prior to the origin of the universe, and yet cause the universe. Again, they try to obviate these issues by characterizing the amplituhedron as being a highly atypical existent which, in a sense, is not an existent, but which causes existents. However, their concept of the amplituhedron is immensely dissociative; and their ensuing mathematical and non-mathematical language about the concept is of meaningless linguistic alteration.

The mathematics of Arkani-Hamed and Trnka does not even minimally conceptually demonstrate that according to the concept of the amplituhedron, it is of pure geometry, contentless, timeless, eternal, non-locational, ahistorical, causal, and non-caused. And of the mathematical language that expresses the above, it is of the array of kinds of meaningless linguistic alterations, just as the above non-mathematical language (the terms) are, except "causal," which of course arise, in the first place, from dissociation. Aside, and as was discussed earlier, mathematics is founded by concepts, and is, as such, always secondary to the concepts. Mathematics, independently, unless founded by accurate concepts, and, or, truly coherent concepts, will not only not be capable of addressing the matters of physics that I discuss, but will provide observations, theory, and statements that are inaccurate, incoherent, meaningless, delusive, dissociative, and psychological in nature.

As an aside, and as is demonstrated above, Arkani-Hamed and Trnka believe that there is a pre-universe realm, that is not spatial nor temporal, which is as such not in the universe, which is where the amplituhedron subsists (which does not have location nor temporality, and instead, only pure geometry), and which is what the universe arose from. Moreover, since they do not argue that this pre-universe realm ceased to subsist upon causing the universe, the pre-universe realm would still subsist. However, they do not provide an even minimal argument, nor an even minimal pictorial representation, of what this pre-universe, non-universe, universe-causing realm is. As was discussed, their conceptual method is of severe multifaceted dissociation; and their ensuing mathematical and non-mathematical linguistics are of meaningless linguistic alteration. Moreover, such concepts and linguistics, due to be combined with an array of other concepts and linguistics that are coherent and meaningful, serve to psychologically captivate.

121. As was discussed earlier, the prevailing general concept of "dark matter" is the following: It is inferred to exist (that is, it is inferred to be what adheres the constituents of galaxies together, thereby preventing the generally integrated constituents of the rotating systems from disintegrating), it is inferred to be a substance, the nature of it is unknown, it has not been observed nor measured, there is no evidence that it interacts with ordinary matter, there is no evidence that it emits light, there is no evidence that it absorbs light, and it is inferred to be unevenly distributed in the observable universe (it is inferred to be clumped, or present to a greater degree, in regions of galaxies). It is moreover referred to as the physical scaffolding on which galaxies develop. However, aside from how there is no explanation about what causes black matter, there is no explanation about what keeps it in place, or, that is, what its scaffolding is. Its scaffolding is the 1-4 dimensional micro external universe. A semblance of black matter and black light is the fifth dimension (the three dimensional surface) of the 1-3|5 dimensional observable universe; and the 1-4 dimensional micro external universe is the mechanism of the observable universe. Hypothetically, without the micro external universe, and setting aside that it is the cause of the fifth dimension, the aforementioned accumulations of the fifth dimension in certain regions of the observable universe (which Randall, et al., refer to as clumps), would not be present. Rather, black matter would be evenly distributed throughout the observable universe; and as such, there would be no galaxies, nor any astronomical objects, except gasses.

As an aside: Since, as was discussed earlier, the macro external universe is also, at least, 1-4 dimensional, and is, as such, connected with the micro external universe, it, also, is a mechanism of the observable universe. In *(36)*, I concluded something similar about black matter, namely that since the black matter of the micro external universe and macro external universe are of the same 1-4 dimensional realm, both are the first cause of, enduring cause of, and mechanism of, the observable universe.

122. It may likewise be asked what holds the micro external universe in place; and to this I will state that the entirety of the external universe is the core structure of the universe.

123. What is the purpose of black light dispersing the content of the observable universe? More generally, what is the purpose of the phenomena and content of the observable universe that is external to earth? While the cause and mechanism of the observable universe is the external universe (the micro external universe and macro external universe), I surmise that the phenomena and content of the observable universe are akin to the multifaceted, interrelated, immensely complex, and still far from fully observed and understood, intricacy of the human body: The phenomena and content of the observable universe surely are intimately interrelated, and dependent on one another, as are the phenomena and content of the human body, such that the general functioning of the observable universe relies on all aspects of its phenomena and content, with, of course, leeway for the occurrence of adverse phenomena and content, damaged phenomena and content, and the loss of phenomena and content.

Notwithstanding that the earth is not located at the center of its solar system, and notwithstanding that its solar system is not located at the center of its galaxy, and notwithstanding that its galaxy is not located in the center of what is currently observed to be the general center of the observable universe, and notwithstanding that it is comparatively diminutive, its nature and rarity, and the nature and rarity of human life and non-human life, indicate that it, and its content, are, nevertheless, radically significant—that is, the most significant facets of the observable universe. And by "significant," I mean that they are of the most immense complexity; and the neurological and non-neurological complexity of humans and non-humans is not even to an extraordinarily minimal extent found anywhere else. As an aside, the significance of a facet of the observable universe is not determined by its location, nor size. Analogously, the brain is not located at a central region of the body, and is not the largest feature of the body. Moreover analogously, if hypothetically the largest feature of the observable universe was a rock, or black hole, or relative ordinary-matter vacuum, etc., and if it existed in the center of the observable universe, this would not entail that it is the most significant facet of the observable universe.

124. Notwithstanding the formal constraints that I discussed previously, I would like to proffer the following possibility regarding the origin of the external universe. I will firstly mention that the prevailing alternatives, as is well known, entail the general problem that if it is argued that X caused, or continues to cause, the universe, the issue of what caused X remains. Moreover, there is the general problem that attempts to argue that X, given X's nature, is uncaused, are simply psychological attempts to obviate the aforementioned problem; and I would add that the psychological attempts typically entail conceptual dissociation, and meaningless linguistic alterations. Now what I would like to proffer begins with a reconception of "nothing" (and synonymously, "zero," "absence," etc). While I discussed in *(17)*, *(18)*, and *(36)* that zero is not something, but rather, the absence of something in particular, and that while it is of course possible to observe, measure, and conceive of the absence of something in particular, there is no such thing as absence itself, because such a purported thing has never been observed nor measured, and because the concept of absence (absence itself) is dissociative, I tacitly conceded that absence (absence itself) can be coherently conceived of (that is, non-dissociatively conceived of). I would now like to conceive of it in the following way: While it is not something, and while, as such, there is nothing to observe and, or, measure, it can be generally conceived of, just as the absence of thought can be generally conceived of, and as death can be generally conceived of. While, of course, the absence of thought, and death, are not states of existence, but rather, the absence of states of existence, and while, of course, both preclude the observations of the people who have encountered them, because the people are no longer present, they can generally be conceived of, by observers, in relation to what I will refer to as "presence." Absence can be conceived of as the absence of presence; and it is of course not accurate to conceive of absence as "the state of absence." Therefore, absence, akin to the absence of thought, and the absence of life, is the absence of presence; and its conception, as such, relies on what is opposite to it, namely the conception of presence.

Now, I will moreover argue that (a) the conception of presence relies on what is opposite to it, namely the conception of absence, (b) presence itself relies on absence itself, (c) absence *converts* to presence; and presence can convert to absence, and (d) absence does not cause presence. To begin: As was discussed, absence is not something; and it, as such, is not of the external universe, nor of the observable universe; and it, moreover, is not in external the universe, nor in the observable universe. Moreover, it is not outside of the external universe; and as such, it does not coexist with the external universe. As an aside, I will add that since it is not of, nor in, the external universe, nor the observable universe, it is formally different than both the external universe and observable universe. Moreover, the disparity is extreme, since it, as I discuss below, is what converted to the external universe: (a) As was discussed about the relation between the observable universe and external universe, the external universe is formally different than the observable universe, due to being outside of the microlimit and macrolimit of the observable universe, and due to being the cause of the observable universe. (b) Since absence is not outside of the external universe, but rather, prior to it, and since it is what the external universe converted from, it is formally different than both the external universe and observable universe. (c) Since the external universe is formally different than the observable universe, the additional formal difference of absence entails that the disparity between it and the observable universe is comparatively extreme. (Aside: Formal difference is already an extreme kind of difference; and since the disparity between the observable universe and absence is of two levels of formal difference, the disparity is comparatively extreme).

As was demonstrated, it is possible to have a general conception of the formally different external universe, namely that it consists of black matter, which emits black light, that it is the cause of the observable universe, and the mechanism of the observable universe, that it is 1-4 dimensional, that the microlimit that it provides for the observable universe is the three-dimensional surface (the fifth dimension) of the observable universe, that the microlimit pervades the entirety of the observable universe, that it forms the macrolimit of the observable universe, and that without the microlimit and macrolimit, there would be blindness for visual beings in conditions of all degrees of 400nm-700nm light, and in the condition of relatively complete darkness. Unlike this general conception of the formally different external universe, it is only possible to minimally generally conceive of absence; and this is consistent with its extreme formal difference.

Only absence (that is, as characterized above) is that which is not subject to the issues of (a) What is the cause of it?, and (b) What is the nature of it such that *(a)* is not an issue?: Only absence cannot have a cause, nor a nature.

There is no microlimit nor macrolimit of absence, because it is not something, nor of the external universe, nor of the observable universe, nor in the external universe, nor in the observable universe. (Aside, I occasionally use the term "it" to refer to absence, rather than repeating "absence," in order to avoid eliciting the confusion that is entailed by implicitly conveying that absence is being discussed anew).

The only possible relationship between absence and the external universe is one of conversion: Absence cannot cause anything; and the only alternative is that of it converting to the external universe.

As is discussed above: Causation, emergence-from, etc., is not, and cannot be, the fundamental origin-mechanism of the universe. Only conversion can be. And as to how precisely the external universe is converted from absence, it is formally impossible to conceive of this. Instead, all that can be done is to state that this is the case. We would have to become an immensely, formally different kind of being in order to be able to possibly conceive of the relation between absence and the external universe.

Causation, therefore, is not the only mechanism for existence. Moreover, causation is never a mechanism for what I will refer to as "original existence": All relative existents and relative phenomena of the observable universe that are observed, measured, and conceived of to be caused by other relative existents and relative phenomena are, actually, simply novel conglomerations, reductions, or reactions of preexisting relative existents and relative phenomena; and the novel relative existents and relative phenomena can simultaneously be partly of conglomeration, partly of reduction, and partly of reaction; and the reactions can of course be internal reactions of the intricacy of relative existents and relative phenomena.

I discuss the following in *(3)*, *(5)*, and *(48)*, regarding the external universe (that is, the micro external universe and the macro external universe); and while the following are my early formulations, they generally apply to my current discussion.

> (a) Since the universe is the entirety of reality, there cannot be anything on the other sides of reality. (b) There surely is an alternative to the states of being microlimited, macrolimited, and unlimited, and that it is something that is formally inaccessible to human experience and conceptualization. (c) Of what exceeds the microlimit and macrolimit of the observable universe, namely the external universe, it is surely neither limited nor unlimited: There surely is an alternative to the states of being limited and unlimited, and that it is something that is formally inaccessible to human experience and conceptualization. (d) Akin to how the external universe cannot have a microlimit, nor macrolimit, nor be unlimited, it also cannot have an origin, nor be "infinite." Moreover, surely there is an alternative to the state of origination, and that it is something that is formally inaccessible to human experience and conceptualization.

The alternative to the external universe being geometrically limited, or geometrically "unlimited," is that it arose via conversion from absence, and could discontinue via conversion to absence: Since it did not arise from a geometric condition, and since it would not discontinue to a geometric condition, it does not have a microlimit, nor macrolimit. It does, though, provide a microlimit and macrolimit for the observable universe, and it does, as such, discontinue at the fifth dimension (the three dimensional surface of the observable universe), which is where the ordinary matter of the observable universe arises. That is, the macro external universe, and the micro external universe, discontinue at the fifth dimension of the observable universe.

Regarding whether there is an alternative to the external universe originating, the external universe does have an origin, but it is not a causal origin. Instead, it is a conversion origin.

As to how absence converts to presence, I believe that the form of the phenomenon is nearly the same as what occurs for the generation of thought. To begin: For most people, their thought is considerably schematic, and considerably autonomic, meaning that it consists of minimal conceptual structures, which rapidly arise and depart; and essentially the entirety of their awareness (their sensory awareness) is directed to their spatial fields. That is, their sensory experiences of the content of their spatial fields comprise essentially the entirety of their awareness; and their thought occurs as minimal conceptual schema which, moreover, rapidly arise and depart, and often are essentially completely supplanted by their sensory experience (and additionally, their emotional experience). It is not the case, for anyone, that their thought consists of conceptual pictorial worlds that are even minimally akin to the vividness and intricacy of the content of their spatial fields. While of course, for everyone, there are relatively rare moments of recollection, envisioning, etc., during which time they experience their thought in nearly the same way that they experience their spatial fields, these moments are relatively rare; and during these moments, they typically have little or no sensory awareness of, nor supervening conceptual reflection on, their spatial fields. Now, it is not the case that people intermittently experience an empty spatial field of thought, and that conceptual schema and pictorializations intermittently arise in it, and intermittently depart from it. That is, people do not experience empty mental mediums in which content intermittently arises and discontinues. Rather, there is absence, and then the arising of conceptual schema and pictorializations, and then the discontinuation of the conceptual schema and pictorializations; and the discontinuation entails the return to absence. And by "absence," I mean what occurs during the above moments when one's thought is essentially completely, or completely, supplanted by sensory experience. While of course there may not be a complete absence of thought, the purpose of this discussion is to demonstrate that during those moments there is no empty mental spatial field.

Thought, then, converts from absence. (Again, thought would not arise *in* a condition of absence, nor *from* a condition of absence, because absence is not a condition). Moreover, since pre-thought absence is not geometric, thought does not arise from a geometric condition. As an aside, thought is a geometric condition; although of course it is not a precise one.

The above demonstrates that a geometric condition can arise via conversion from non-geometric absence. The form of this phenomenon is the same as the form of the phenomenon of the external universe converting from absence: The 1-4 dimensional geometric condition, which provides the fifth dimension of the 1-3|5 dimensional observable universe, arose via conversion from non-geometric absence.

Absence converts to thought; and when thought discontinues, there is absence.

Of course it is the case that without the brain, there would not be thought; and of course it is the case that, as such, the brain causes thought. However, the precise mechanism of cause between the brain and thought is not only unknown, but surely formal unobservable, unmeasurable, and inconceivable. The purpose of my discussion is to describe the conversion of absence to thought, and thought to absence, regardless of the general cause of thought.

Regardless of the general cause of thought, I would argue that if there was no absence for thought to convert from, and if thought could not revert to absence, there would be no thought.

Of course, the aforementioned phenomenon is that of the conversion of absence to non-physical thought, and the conversion of non-physical thought to absence, and that an analogy was made between this and the conversion of absence to the external universe, and the conversion of the external universe to absence. While the analogy of course does not demonstrate that the conversion occurs for the non-mental universe, it does provide a demonstration of the general form of the phenomenon. Moreover: Without the analogy, the conversion would remain of theory: The analogy provides an account of an actual occurrence of conversion.

Now, regarding the relation between absence and presence, and in particular, the relation between absence and thought, and the relation between absence and the external universe, there are the issues of (a) why absence converts to thought, (b) why thought is as it is, (c) why absence converts to the external universe, and (d) why the universe is as it is. To begin: Regarding *(b)*: As was discussed, absence does convert to thought, in the context of the brain. While what can be thought can of course exceed, in nature, and extent, what is found in the observable universe, this is nevertheless done via mental modification of what is found in the observable universe: Hypothetically, without the observable universe, thought would not convert from absence. Absence would remain, despite the capacity for thought. Now regarding *(a)* (why absence converts to thought), one's sensory experience, combined with one's psychology, compel the conversion: Absence changes into a semblance of one's sensory experience and one's psychology, and only a semblance, which entails that it does not dominate the course of one's life: As was discussed, one's thought occurs as a minimal schemata; and this entails that one's sensory experience, psychology, and conceptualization is predominately, and often almost entirely, directed to the content of one's spatial-field. (Aside: For persons with marked autism-spectrum-disorder, the opposite is the case: Their thought is predominately of maximal schemata; and this entails that the content of their spatial-fields is predominately minimally experienced, with the exception of sporadic moments during which it is experience to a comparatively extreme degree, which entails that their thought is either, depending on how intensely that they experience their spatial-fields, of comparatively extremely minimal schemata, or essentially non-existent, or non-existent).

Regarding *(c)* (why absence converts to the external universe): Unlike what is the case for thought, there is not anything that compels absence to convert to the external universe. And regarding *(d)* (why the universe is as it is): Unlike what is the case for thought, the external universe is not a schemata of something else. Moreover, and as such, the external universe does not rely on something else for absence to convert to it.

Regarding *(c)* (again, why absence converts to the external universe): To begin: (a) I demonstrated that absence can convert to something, and that something can revert-convert to absence, and that the conversion and reversion-conversion can occur persistently. This demonstrates that these phenomena are possible. If thought consisted of a perpetual spatial-realm within which schema persistently arose and discontinued, there would not be absence in that context, and, as such, I would have not been able to demonstrate the above. (b) There are no other absence conversion phenomena, nor absence-reversion conversion phenomena, of the observable universe. (c) Without the absence of thought, there would not be demonstrations of, nor, as such, bases to conceive of, absence. Instead, there would only be relative absence, and conceptions thereof. Aside, while it would be common, as it is currently, to believe, and state, that particular contexts consist of nothing (or are absent of anything), those contexts are only of relative absence. (d) Absence cannot be permanent, because absence of thought is not permanent, and because there is an observable universe and external universe. (e) Absence is such that it converts to what is opposite to it, namely, again, presence. (f) Presence is such that it can revert-convert to absence. (g) Again, while it is commonly believed, and stated, that "space" is absent of anything, and that other realms of the observable universe, such as the "space" between the relative sub-existents of relative micro-existents, consist of this empty space, the microlimit of the observable universe (black matter) pervades the entirety of the observable universe. As such, presence consists entirely of presence; and as such, the conversion from absence to presence is complete. And this is the case for thought as well: Thought consists entirely of thought; and the conversion from absence to thought is complete. (h) Since there is not anything that compels the conversion from absence to the external universe, what remains is to derive what can be derived of absence, namely the following: It is radically formally different than both the observable universe and the external universe; and as such, predominately only what it is not can be conceived of, namely that it is not an existent, nor a state of existence, nor physical, nor mental, nor causal, nor caused. As for what it is, all that can be conceived of is that it converts to the external universe and thought, and can be reverted-to via the conversion of thought to it, and likely via the conversion of the universe to it. And as for the mechanism of the conversion between absence and thought, and absence and the universe, it also is radically formally different, and as such, formally inconceivable for humans. I will, however, provide two additional derivations: First, the above about absence is the case because it is what I will refer to as "the most simple"; and I only state, "the most simple," because, as was discussed, it is not the most simple "feature," and because it is not the most simple "feature" of "the universe." And as for what I mean by "the most simple," I mean that it, unlike even the general form of the external universe, cannot be delineated in terms of form, nor in terms of mechanism: That is, it is not an existent, nor a state of existence, nor a condition; and it is not physical, nor mental; and it is not caused, nor causal. Second: As was discussed, absence converts, and can be reverted to; and it converts to what is obverse to it, namely presence; and it can be reverted to from what is obverse to it, namely, again, presence. I will argue that presence cannot be anything other than the external universe and observable universe. To begin: While the formal aspects of the observable universe and external universe that I delineated are the case, what we observe, measure, and conceive of as the intricacy of the observable universe is, according to my discussion about the formal aspects of the observable universe and external universe, not even minimally the case. In particular: There are no discrete existents; the microlimit (black matter) pervades the observable universe; the microlimit, and its relation to ordinary matter, is formally different than anything of the observable universe, and as such, cannot be observed, measured, nor precisely conceived of; what ordinary matter is, itself, cannot be observed, measured, nor precisely conceived of; and what we observe, measure, and conceive of, of the observable universe, is completely dependent on our neurologies (our sensory faculties), including the sizes of their apparatuses. Now, the issue of why the content of the observable universe is as it is, is a persistent issue; yet since the above is the case about the observable universe, the question cannot be answered: It is necessary to know what precisely something is before we can answer why it is as it is. As such, the observable universe, aside from being accurately conceived of as 1-3|5 dimensional, and as being inexorably imbued with black light, etc., it, for the purpose of this particular discussion, can only accurately be conceived of as consisting of unevenly distributed, contiguous, physical substance (and likely the substance is relatively only slightly unevenly distributed), which, moreover, is different in nature throughout its distribution, and perhaps relatively only slightly different. Moreover as such: Notwithstanding that the external universe is 1-4 dimensional, and, moreover, formally different than the observable universe, etc., it, also, for the purpose of this particular discussion, can only accurately be conceived of in the above way. The question of what presence is, then, is a question of why there are 1-4 dimensional and 1-3|5 dimensional realms, and unevenly distributed physical substance, and physical substance of different natures throughout the distribution, and, of course, physical substance itself. Aside: As is entailed in my discussion about condensed space, I consider the various "forces," and the like, to actually be aspects of the substance of the external universe and observable universe—that is, that they are substances themselves—rather than how they are conceived of, namely as "forces," which I consider to be a dissociative concept that arises from the limitation of observation and measurement: What is firstly conceived of is that relatively discrete objects can directly interact in such a way that they pull, push, collide, combine, disintegrate, etc; and then what is conceived of is the phenomena of pushing, pulling, colliding, combining, disintegration, etc., without the direct object-interaction: The object-interaction is dissociated from, and it is then delusively believed that the above phenomena are phenomena that are separate from direct object-interaction.

I will argue that the opposite of absence is, and can only be, physical substance, and likewise, that presence is, and can only be, physical substance. Moreover: As for the most general aspects of the external universe and observable universe, and why they are as they are, I believe that all that can be conceived of is that absence converts in this way: Since absence is radically formally different, it, and its mechanism of conversion, cannot be conceived of beyond how I conceived of them.

Regarding the possible nature of the physical substance: To begin: I believe that the following occurred at the instant of the conversion of absence to it: The substance appeared as one, evenly distributed, evenly dense, contiguous substance that was located in the relatively widespread 1-4 dimensional region that comprises the external universe. Thereafter, the substance interacted with itself such that the current external universe was formed. Thereafter, the external universe simultaneously caused (and still causes), and served as the mechanism for (and still serves as the mechanism for), the observable universe. Now as for the substance itself, it can only be derived as being the first level of a three level substance, namely it, that of the external universe, and that of the observable universe), none of which can be observed, measured, nor conceived of for what they precisely are. Moreover, while it would have been formally the same as that of the external universe, it would have been immensely combinatorially different.

Now as for the issue of why what I will refer to as the "initial external universe" was not an inexorably static substance, I believe that there is no such substance, and never could be: All physical substance, especially at different micro-depths, intensely interacts with itself.

As for the issue of why the observable universe is as it is, it would be necessary to observe, measure, and conceive of the substance of the initial external universe, the substance of the external universe, and the substance of the observable universe, in order to be able to possibly accurately conceive of why. Moreover, and as was discussed earlier, we would have to be formally different than we are in order to be able to observe, measure, and conceive of the substance of the initial external universe, the substance of the external universe, and the substance of the observable universe, and in order to ensuingly be able to possibly accurately conceive of why the observable universe is as it is. That is, we would have to not be in the observable universe, nor of the observable universe; and instead, we would have to be in the external universe, and of the external universe.

As for the "size" of the external universe: As was discussed earlier, it is not the case, as is the case of the size of the observable universe, that it is formally impossible for us to observe, measure, and conceive of the size of the external universe, but rather, that the external universe is not microlimited, nor macrolimited, nor unlimited. As such, there is no issue of size. I will though provide the following additional discussion. I previously stated that there surely is an alternative to the external universe being microlimited, macrolimited, and unlimited; but I did not provide an alternative. Based on my discussions between that time and the present, I will argue that the external universe does have a microlimit and macrolimit, and that they are of the sixth spatial dimension, which I will refer to as "outward"; and recall that the fourth spatial dimension is inward. The sixth dimension, like the fourth dimension, comprises both a microlimit and macrolimit; although unlike the fourth dimension, the traversing of it does not result in the entry into another dimensional realm, but rather, the conversion to absence.

Absence, therefore, does not fully convert to presence.

The external universe can therefore be conceived of as 1-4|6 dimensional.

Regarding the time that would have elapsed (that is, the clock-time) for the conversion of absence to the initial condition of the external universe, the conversion would have been instantaneous: No time would have elapsed.

Regarding the emergence of relatively separate biological existents: Given that the substance of the observable universe, external universe, and initial condition of the external universe, cannot be observed, measured, and accurately conceived of, it is not possible to derive how precisely any of the relative existents are as they are: Again, it is necessary to firstly observe, measure, and accurately conceive what they are before it can be determined how they emerged. I believe that all that can be accurately conceived of is that they emerged from unique combinations of the substance of the observable universe. As an aside: Given that there is a relatively extreme rarity of biological existents, and other facets of the earth, and a relatively extreme abundance of other observable substances and phenomena in the observable universe, I believe that the earth and its content emerged from profoundly random combinations of the substance of the observable universe. That is, I believe that the content of the observable universe emerged from random combinations of the substance of the observable universe, and that the extremely rare content of the observable universe emerged from combinations that are comparatively extremely random.

I would now like to modify some earlier discussions in light of my discussions about the initial substance of the external universe, the later substance of the external universe, the substance of the observable universe, and that the initial substance of the external universe was converted from absence. The initial substance and geometrical system that was converted from absence could have been both the substance and geometrical system of the 1-4|6 external universe, and the substance and geometrical system of the 1-3|5 observable universe. As such, both of the formally different systems would have had initial states, which then changed via the interaction of the substances with themselves, and with each other. I believe that this is an increasingly more accurate conception, as my previous conception that the external universe caused the observable universe did not include a causal explanation, and attributed the cause to formally different aspects of the external universe. While the concept of cause as conversion does not include a concept of the mechanism of the conversion, it is, notwithstanding, specific, and, I would argue, immensely superior to the prevailing concept of cause as interaction: No true cause is actually observed, measured, and conceived of, but rather, only the results of the interactions of pre-existing relative existents and phenomena.

I would moreover like to provide the following modification. To begin: As was discussed, the external universe (the micro external universe, and the macro external universe) has a microlimit and macrolimit, namely the sixth spatial dimension, which I refer to as "outward." Traversing those limits entails traversing to absence. Now, as is the case of the observable universe, the microlimit of the 1-4 dimensional external universe would be a three dimensional surface; and as such, both the observable universe and external universe have the fifth spatial dimension. (Aside, just as the macro limit of the observable universe is comprised of the micro limit of the observable universe, the macro limit of the external universe is comprised of the micro limit of the external universe; and as such, the macro limit of the external universe is also of the fifth spatial dimension). As such, the external universe is 1-4|5 dimensional. It is unclear, though, what the fifth spatial dimension of the external universe is comprised of, unlike what is the case for the observable universe: The fifth dimension of the observable universe is comprised of black matter. Notwithstanding, traversing the fifth dimension of the external universe would result in the conversion to absence. That is, the traverse of the fifth dimension of the microlimit and macrolimit of the external universe is a traverse to absence. The external universe is therefore 1-4|5|6 dimensional. Aside, the observable universe, which is 1-3|5 dimensional, is not 1-3|5|4 dimensional, because traversing its fifth dimension entails entry into the external universe via the external universe's fourth dimension (the dimension of inward). Traversing the fifth dimension of the external universe traverses to absence, via the sixth dimension (the dimension of outward); and since the sixth dimension is not a dimension of a spatial realm, it can be considered to be an aspect of the dimensional system of the external universe.

I would moreover like to provide the following modification. Given that the micro limit and macro limit of the observable universe are of the fifth spatial dimension (that is, of three-dimensional surface), and that while they are *in* the observable universe, they are *of* the external universe, and that they are, of course, of pure black matter, the immediate black light that they emit is of pure black light. The black light, then (that is, farther away from black matter), interacts with 400nm-700nm light, as was discussed. As such, not all of the black light of the observable universe is non-pure. And this is consistent with my previous argument that ideal super-resolution microscopy would show that black light pervades the entirety of the observable universe at the microlimit and macrolimit of the observable universe, including in all conditions of 400nm-700nm light, and including on the relative surfaces of all 400nm-700nm light sources themselves. The microscopy would, in a sense, look beyond the 400nm-700nm light, to black light.

125. Regarding the thirty-eight principles of optical geometric physics in *(89)* and *(116)*, the following are modifications of some of the principles, replacements of some of the principles, and new principles.

(39) The scaffolding of the micro limit of the observable universe and macro limit of the observable universe—that is, the scaffolding of the black matter in the observable universe—is the micro external universe and macro external universe.

(40) The micro external universe and macro external universe—that is, the entirety of the external universe—is the core structure of the universe; and as such, it does not have a scaffolding.

(41) The external universe is 1-4|5|6 dimensional. The fourth dimension is the dimension of inward. The external universe is located within the observable universe via the dimension of inward. The fifth dimension is the three dimensional surface microlimit and macrolimit. At any location of the fifth dimension, the sixth dimension of outward can be accessed. The sixth dimension traverses to absence.

(42) The origin of the universe occurred via the conversion of absence to the universe. The universe may then revert-convert to absence; and thereafter, absence may convert again to a universe.

Absence converts to what is opposite to it. There are two kinds of absence: absence of the mind, and absence of the universe. For the mind, absence converts to thought; and thought is the opposite of absence of the mind. For the universe, absence converts to the universe; and the universe is the opposite of absence of the universe.

126. An addition to *(108)*. An article by Einstein, Podolsky, and Rosen generally characterizes the theory of quantum entanglement, and provides their objection that the theory is incomplete. They argue that "the operators [or physical systems] corresponding to "two physical quantities" that "do not commute [communicate]" "cannot have simultaneous reality": The physical systems do not communicate because (a) "the process of measurement carried out on the first system" "does not disturb the second system in any way," and (b) the physical quantities of the two systems cannot be "simultaneously measured or predicted."

> Previously we proved that either (1) the quantum-mechanical description of reality given by the wave function is not complete or (2) when the operators corresponding to two physical quantities do not commute[,] the two quantities cannot have simultaneous reality. Starting then with the assumption that the wave function does give a complete description of the physical reality, we arrived at the conclusion that two physical quantities, with noncommuting operators, can have simultaneous reality. Thus the negation of (1) leads to the negation of the only other alternative (2). We are thus forced to conclude that the quantum-mechanical description of physical reality given by wave functions is not complete.
>
> One could object to this conclusion on the grounds that our criterion of reality is not sufficiently restrictive. Indeed, one would not arrive at our conclusion if one insisted that two or more physical quantities can be regarded as simultaneous elements of reality *only when they can be simultaneously*

> *measured or predicted.* On this point of view, since either one or the other, but not both simultaneously, of the quantities P and Q can be predicted, they are not simultaneously real. This makes the reality of P and Q depend upon the process of measurement carried out on the first system, which does not disturb the second system in any way. No reasonable definition of reality could be expected to permit this.[58]

Notwithstanding their objections, the prevailing interpretation of the phenomena of quantum entanglement is that the mechanism of communication between once entangled (and therefore communicating) physical systems, but now entirely separate systems, must be determined. However: As I demonstrated, there are no discrete existents, but rather, only relative existents; and as such, once entangled physical systems are never separate, and can only become separate when their entanglements are replaced by new entanglements with other physical systems. More specifically, the seemingly distance-separate relative existents that are determined to still be entangled have only a facet of themselves measured: They extend across the distance between the facets of themselves that are measured. And since the extension is currently not observable, it is an extension of one or more novel substances.

127. 400nm-700nm light, of particular intensities, emitted via particular methods, and emitted in particular environmental conditions, can appear in the air, and over the surfaces of various materials, and within various materials, as generally white, and as generally red, orange, yellow, green, blue, and violet. However, the light, itself, is invisible, as was discussed: The composition of the air, and the composition of the surfaces of the materials, and the composition within the materials, are

---

58. Albert Einstein, Boris Podolsky, Nathan Rosen, "Can Quantum-Mechanical Description of Physical Reality Be Considered Complete?," *Physical Review*, Volume 45, https://journals.aps.org/pr/pdf/10.1103/PhysRev.47.777, May 15, 1935.

such that general portions of the invisible light are reflected as particular colors. Moreover, the experience of the reflections as being particular colors is completely dependent on one's neurology, which of course including one's visual system. While it is objectively the case that 400nm-700nm light consists of different sub-types of light, it is not objectively the case that the sub-types, when they are reflected on particular substances, are of particular colors.

128. Regarding possible advancements in super-resolution microscopy: Based on what has been discussed to this point, the following is the case: (a) It could never be determined that a particular resolution is the final resolution—that is, that there are no more powerful resolutions. (b) Of whatever is observed via a presumed final resolution, it, by virtue of being observed, could be observed by more powerful resolutions. (c) It is only when a resolution enters a state of black light that cannot be resolved with 400nm-700nm light that it is possibly the case that the resolution has exceeded all of the observable relative existents of the observable universe, and has entered the realm of black matter and pure black light. (d) Regarding *(c)*: In order to determine whether a resolution has entered the realm of black matter and pure black light, it would however be necessary to exceed the microlimit of the observable universe, as it is only from that formally different perspective, and by observational beings who are formally different than we are, namely at least 1-4|5|6 dimensional, could it be observed and, or, measured that the resolution has entered the realm of the microlimit of the observable universe. (e) Regarding *(d)*: There would however be increasing certainty that the microlimit of the observable universe was entered if, perpetually, each more powerful resolution, regardless of the extent of 400nm-700nm illumination, only observed black light.

129: Conclusion. While the precise content of the external universe and observable universe cannot be observed, measured, and conceived of, and while the fourth dimension of the 1-4|5|6 dimensional external universe cannot be entered by us from the 1-3|5 dimensional observable universe, and while the sixth dimension of the 1-4|5|6 dimensional external universe, which traverses to absence, cannot as such be traversed by us, we can observe black light, and we may be able to measure via microscopy, or one or more other methods, that black light pervades the entirety of the observable universe, including in all conditions of 400nm-700nm light, and including on the relative surfaces of all 400nm-700nm light sources themselves, which entails that the fundamental color of the entirety of the content of the observable universe is black, and that the microlimit of the observable universe, which pervades the fifth dimension of the 1-3|5 dimensional observable universe, but which is of the 1-4|5|6 dimensional external universe, is black matter, and emits black light. Moreover, we can observe the macrolimit of the observable universe via its black light. Moreover, we know that gravity is caused by condensed microlimit, and the micro external universe (the 1-4 dimensional mechanism of the observable universe); and we know that the microlimit (space, which pervades the entirety of the observable universe) cannot be curved. And we know that a considerable extent of fundamental concepts of physics and mathematics are dissociative, and ensuingly of the many variations of meaningless linguistic alterations. And we know that the mechanism of the observable universe is the external universe, and that it, like the black light that it emits, is formally different than the observable universe. And we know that we, with our neurologies, and capabilities, are only able to observe, measure, and conceive of a semblance of the observable universe, and nothing of the external universe, except the black light that it emits into the observable universe. And we can derive the existence of the microlimit and macrolimit of the observable universe via the optics of black light, and via the geometric analyses that I provided. And we know that the observable universe is not fundamentally comprised of particles, nor any other kind of relative existent, and that it, instead, is of one contiguous substance. And we know that the external universe and observable universe were not caused via the modification of one or more preexisting relative existents, phenomena, or states, but rather, that they were converted from absence. And while we can only have extremely general concepts of absence, and of the conversion of absence to presence, and mostly by way of the analogy of the conversion process of the mind, we do know the form of what occurs.

The form of the content of the universe is the following: Absence; the mechanistic content of the external universe; black matter in the observable universe, which is *of* the external universe, and which is formally different in nature than the content of the observable universe, and which is of a different physics; the black light in the observable universe, which is of the external universe, and which is formally different in nature than the light of the observable universe, and which is of a different physics; and the ordinary matter of the observable universe.

The form of the structure of the universe is the following: Absence; the 1-4|5|6 external universe; and the 1-3|5 observable universe.

There are formal constraints that preclude us from observing, measuring, and conceiving of the precise fundamental content and precise fundamental structure of the external universe and observable universe; but via my geometric analyses, the observation of a semblance of black matter, the observation of black light, my optical geometric analyses, optical analyses, mathematical analyses, and conceptualization, we can conceive of their forms. We can, though, observe and conceive of a facet of the precise content of the external universe and observable universe via the observation of a semblance of black matter, and via the observation of non-pure black light. Moreover, we can observe and conceive of the 1-3|5 spatial dimensionality of the observable universe.

My initial geometric analyses are confirmed by what I later derive from my observation of a semblance of black matter and of black light.

If we could observe, measure, and conceive of the precise fundamental nature of the external universe and observable universe, we would not be as we are, nor where we are.

References

Aguirre, Anthony, "What Would an Infinite Universe Mean?" *Closer To Truth,* www.closertotruth.com/interviews/1684, November 2014.

Ananthaswamy, Anil, "The New Shape of Reality," *New Scientist,* 1377, p. 1. https://trnka.faculty.ucdavis.edu/wp-content/uploads/sites/322/2016/02/new-scientist.pdf, July 29, 2017.

Arkani-Hamed, Nima, "The Future of Fundamental Physics," *Daedalus,* Journal of the American Academy of Arts & Sciences, p. 66, www.amacad.org/publication/future-fundamental-physics, www.sns.ias.edu/ckfinder/userfiles/files/daed_a_00161(1).pdf, Summer 2012.

———, "What are the Ultimate Questions of Nature?" 4:00-4:20, *Closer To Truth,* www.closertotruth.com/interviews/1709, 2016.

Arkani-Hamed, Nima and Jaroslav Trnka, "The Amplituhedron," *Journal of High Energy Physics,* https://arxiv.org/pdf/1312.2007.pdf, Dec. 6, 2013.

CajaCanarias Foundation, Interview with Lisa Randall, Video: "Turning On the Cosmos," 4:40-4:52, https://youtu.be/juZ5nfvhhCU, 2015.

Carroll, Sean, "Dark Photons," *Discover Magazine,* www.discovermagazine.com/the-sciences/dark-photons, October 30, 2008.

———, "The Preposterous Universe," www.preposterousuniverse.com/preposterous.html, 2008.

CERN, "Dark Matter," www.home.cern/science/physics/dark-matter, 2020.

CERN-CMS, "No Sign Of Dark Light From The Higgs Boson," https://cms.cern/news/no-sign-dark-light-higgs-boson, 2016.

Davidson, Michael W., Andreas Nolte, and Lutz Höring, "Fundamentals of Illumination Sources for Optical Microscopy," Zeiss, http://zeiss-campus.magnet.fsu.edu/articles/lightsources/lightsourcefundamentals.html.

Davidson, Michael W., and Kenneth R. Spring, "Image Brightness," MicroscopyU, www.microscopyu.com/microscopy-basics/image-brightness.

Einstein, Albert, Boris Podolsky, and Nathan Rosen, "Can Quantum-Mechanical Description of Physical Reality Be Considered Complete?," *Physical Review,* Volume 45, https://journals.aps.org/pr/pdf/10.1103/PhysRev.47.777, May 15, 1935.

European Space Agency, "First 3D Map of the Universe's Dark Matter Scaffolding," July 2007.

Hawking, Stephen, *A Brief History of Time,* Bantam Books, 1998.

Institute for Advanced Study, "Nima Arkani-Hamed on the Amplituhedron," www.ias.edu/ideas/nima-arkani-hamed-amplituhedron, 2015.

———, Article: "The Smallest Particles: What Do They Reveal?," https://www.ias.edu/ideas/2013/dijkgraaf-particles-video, November 16, 2013.

———, Video: "The Smallest Particles: What Do They Reveal?," 50:50-51:20, https://www.ias.edu/ideas/2013/dijkgraaf-particles-video, November 16, 2013.

Kuhn, Robert Lawrence, "Space, Confronting the Multiverse: What 'Infinite Universes' Would Mean," www.space.com/31465-is-our-universe-just-one-of-many-in-a-multiverse.html, December 23, 2015.

Linde Andrei, "How Many Universes Exist?" *Closer To Truth,* www.closertotruth.com/interviews/2606, December 2015.

Lopes, Ana, "CMS Hunts For Dark Photons Coming From The Higgs Boson," CERN, https://home.cern/news/news/physics/cms-hunts-dark-photons-coming-higgs-boson, May 24, 2019.

———, "NA64 Casts Light On Dark Photons," CERN, https://home.cern/news/news/physics/na64-casts-light-dark-photons, July 22, 2019.

Massey, Richard, et al."Dark Matter Maps Reveal Cosmic Scaffolding," *Nature,* January 2007.

NASA, "A Tour of the Dark Energy Quasar Survey," Chandra X-Ray Observatory, Harvard-Smithsonian Center for Astrophysics, www.chandra.harvard.edu/resources/podcasts/ts/ts290119_hd.html, February 15, 2019.

———, "Dark Energy, Dark Matter," https://science.nasa.gov/astrophysics/focus-areas/what-is-dark-energy, January 2020.

———, "Quantum Foam," https://science.nasa.gov/science-news/science-at-nasa/2015/31dec_quantumfoam, December 2015.

Palus, Shannon, "Amplituhedron May Shape the Future of Physics," *Discover Magazine,* www.discovermagazine.com/the-sciences/amplituhedron-may-shape-the-future-of-physics, January 7, 2014.

Randall, Lisa, "How is the Cosmos Constructed?," *Closer To Truth,* www.closertotruth.com/interviews/2876, 2017.

———, "Are There Extra Dimensions?," *Closer To Truth,* www.closertotruth.com/interviews/2874, 2017.

Rose, Charlie, Interview with Lisa Randall, www.charlierose.com/videos/23185, 2015.

Schwarzschild, Bertram, "Three-Dimensional Mapping of Dark Matter Reveals the Expected Filamentary Scaffold," *Physics Today,* March 2007.

Susskind, Leonard, "How Do Particles Explain the Cosmos?," *Closer To Truth,* www.closertotruth.com/interviews/3078, 2016.

Wolchover, Natalie, "A Different Kind of Theory of Everything," *The New Yorker,* www.newyorker.com/science/elements/a-different-kind-of-theory-of-everything, February 19, 2019.

———, "A Jewel at the Heart of Quantum Physics," *Quanta Magazine,* www.quantamagazine.org/physicists-discover-geometry-underlying-particle-physics-20130917/, September 17, 2013.

———, "Cosmic Triangles Open a Window to the Origin of Time," *Quanta Magazine,* www.quantamagazine.org/the-origin-of-time-bootstrapped-from-fundamental-symmetries-20191029/, October 29, 2019.